PLEASE STAMP DATE DUE, BOTH BELOW AND ON CARD

DATE DUE	DATE DUE	DATE DUE	DATE DUE
JUN 14 2002			
3-31-03			

Solid–Liquid Interface Theory

ACS SYMPOSIUM SERIES **789**

Solid–Liquid Interface Theory

J. Woods Halley, EDITOR
University of Minnesota

American Chemical Society, Washington, DC

Library of Congress Cataloging-in-Publication Data

Solid–Liquid Interface Theory / J. Woods Halley, editor.

    p. cm.—(ACS symposium series ; 789)

Developed from a symposium sponsored by the Division of Colloid and Surface Chemistry at the 218th National Meeting of the American Chemical Society, New Orleans, Louisiana, August 22–26, 1999.

Includes bibliographical references and index.

ISBN 0–8412–3717–4

    1. Solid–liquid interfaces—Congresses.

    I. Halley, J. Woods (James Woods), 1938- II. American Chemical Society, Division of Colloid and Surface Chemistry. III. American Chemical Society, Meeting (218th : 1999 : New Orleans, La.) IV. Series.

QD509.S65 S66   2001
541.3′3—dc21   00–67637

Foreword

The ACS Symposium Series was first published in 1974 to provide a mechanism for publishing symposia quickly in book form. The purpose of the series is to publish timely, comprehensive books developed from ACS sponsored symposia based on current scientific research. Occasionally, books are developed from symposia sponsored by other organizations when the topic is of keen interest to the chemistry audience.

Before agreeing to publish a book, the proposed table of contents is reviewed for appropriate and comprehensive coverage and for interest to the audience. Some papers may be excluded to better focus the book; others may be added to provide comprehensiveness. When appropriate, overview or introductory chapters are added. Drafts of chapters are peer-reviewed prior to final acceptance or rejection, and manuscripts are prepared in camera-ready format.

As a rule, only original research papers and original review papers are included in the volumes. Verbatim reproductions of previously published papers are not accepted.

ACS Books Department

Contents

Organic Liquid–Solid Interfaces

Indexes

Preface

For two centuries, theoretical models of solid-liquid interfaces have played an important role in their study, highlighted, for example, by Faraday's identification of the passivation layer (*1*), the Gouy–Chapman model of the double layer (*2*), and Marcus' models of electron transfer (*3*). Despite successes, theory lacks first principles predictive power in electrochemistry and related problems of solid–liquid interaction. Even qualitative atomistic and electronic descriptions of common phenomena are still controversial or lacking. In the past decade a combination of circumstances has begun to improve this situation and progress toward a quantitative theory of electrochemical phenomena is within reach, at least for some of the simpler cases. This volume describes aspects of this recent progress by many of the leading research groups.

Consider, for example, the differential capacitance of the electrode electrolyte interface in the "double layer" region of potential, in which no faradaic reactions involving electron transfer are taking place. This quantity was recognized very early, with the work of Gouy and Chapman (*2*) as well as of Helmholtz (*4*) in the late nineteenth century, to be an indicator of the distribution of charges near the interface. Only in the mid-nineteen-eighties, however, did workers begin to take seriously the idea that charge rearrangements taking place at the interface in this region are significantly affected by the quantum mechanics of the electrons of the metal electrode. And only still more recently have quantitative models of this effect been able to describe experiments at all accurately. (*See* Chapter 2 by Walbran and Halley in this volume as well as reference 5 for more details.) Even here, the treatment of transition metal-electrolyte interfaces using first principles Car–Parrinello-like methods (*6*) remains beyond reach.

At another level of experimental complexity, one would like to model faradaic reactions at electrochemical interfaces. This is fundamentally a non-equilibrium kinetic problem, but even its equilibrium aspect, the calculation of the electrode potentials at which the oxidized and reduced species are in thermodynamic equilibrium is a theoretically difficult one. The problem is that even in the relatively simple case of outer sphere electron transfer reactions, equilibrium potentials are differences of large energies associated with the ionization potentials, solvation energies, work functions, and points of zero charge of the reacting ions and the electrode. All of these items must be calculated to an accuracy of less than a tenth of a volt to make useful predictions, but the constituent terms can be as large as 10 volts or more. This problem is discussed in reference 7, for example. One is now able to "postdict" some equilibrium potentials for the simplest reactions with some degree of confidence but the calculation of equilibrium potentials is still not possible for many reactions.

A really useful theoretical electrochemistry must be able to calculate rates of reaction. That is, it must solve the kinetic as well as the equilibrium problem. Here some quantitative progress has been made with the simplest "outer sphere" electron transfer reactions in which the redox center never gets close enough to the electrode to significantly modify the electronic structure of either the electrode or of the reactant. This progress has been achieved by piecing together results from molecular dynamics, statistical mechanics, and first principles electronic structure calculation (8). Results are in reasonable agreement with experiment in some respects, but some qualitative features are still not really right. For example, in reference 8 on the cuprous-cupric reaction rate at a copper electrode, the calculated transfer coefficient, which describes the dependence of the rate on the over-potential near the equilibrium point, disagreed with experiment by roughly a factor of three. Furthermore, though calculated free energy barriers were quantitatively in reasonable agreement with experiment, the calculated prefactors did not agree with experiment to within several orders of magnitude. Whether this was because transition state theory was used in the calculation or for other reasons is not fully understood.

Calculation of rates of so-called "inner shell" reactions, including electron transfer, dissolution, and adsorption, has not proceeded very far. In the present volume, Chapter 6 by B. B. Smith, Chapter 5 by M. Philpott et al., and Chapter 4 by W. Schmickler address aspects of this kinetic problem. Smith reports a variety of calculations on aspects of electron transfer at semiconductor-electrolyte interfaces in which the dynamics of the charge carriers in the semiconductor are extremely important (in contrast to the metal-electrolyte interface). These extensive calculations are giving extremely interesting results and new qualitative insights but are not yet at a level that permits quantitative prediction. M. Philpott et al. are attempting calculation of dissolution rates at metal-water interfaces. This research is a pioneering effort, not yet subjected to quantitative experimental tests. W. Schmickler points out the continued useful-ness of the somewhat schematic Newns–Anderson model for understanding the essential features of reactions at the electrode-electrolyte interface.

Ultimately, for electrochemical applications as well as in other liquid-solid interfaces, theory must confront the realities that electrodes are not smooth, undergo reconstructions as a function of over potential, and are often covered with oxides or adsorbants. Most of these problems have only received very preliminary treatments, even in qualitative models. Roughness has traditionally been modeled in extremely simple ways. Only a few connections have been made with the extensive statistical mechanical literature on roughening transitions at model interfaces (9). Some attempts to extend the existing phenomenological models of roughness in interpretation of capacitance data are in a preliminary stage (10). We are not aware of any first principles attempt to account for reconstruction (11). In the present volume, several authors have contributed work on the problem of modeling oxides at liquid–solid interfaces. Each of these works attacks a part of the problem, but a full model containing all the key features is not yet in hand. For example, the models of Rustad and Felmy in Chapter 9 contain the essential feature of water dissolution at oxide interfaces, but do not take full account of changes in covalency at the interface, whereas the tight binding models described by Schelling and Halley in Chapter 11 have not yet incorporated the electrolyte at the interface.

x

Thus the field of theoretical liquid–solid chemistry and physics has a clear program. It also has new tools. Most notably, these are vastly increased computational power, and improved algorithms to take advantage of it and thus link calculated microscopic structures and dynamics to observations usually involving larger time and length scales. Nevertheless the challenges remain daunting both computationally and conceptually, requiring mastery at both levels of the solid and liquid states and their interactions as well as of the mix of classical and quantum physics that is involved.

Mere extension of existing techniques to larger systems will not, however, lead to the full predictive theoretical capability from first principles which is the ultimate goal. This is not because qualitative insight is sometimes lost when computational models replace analytical ones but also because simple estimates show that even if the unlimited applicability of Moore's law (*12*) were assumed for the future of computational capability, the requirements of present first principles calculations are such that many electrochemical and other liquid-solid phenomena would resist full description for centuries. This aspect of the problem is common to all condensed matter simulation. It is the reason for emphasis on order N methods (*13*), which have only been applied to liquid–solid interfaces in a preliminary way.

A second direction for resolution of the computational challenge lies in systematic coupling of first principles calculations at short scales to calculations at successively larger scales. I have addressed this issue in some detail else-where (*5*). One seeks a way to pass from first principles through molecular dynamics to an appropriate macroscopic theory such as hydrodynamics in a systematic, causal way, so that the outputs of the calculations on the small scales fix the parameters and boundary conditions at the next scale up in the hierarchy. This is being done in piecemeal fashion in some of the work described here. The simplest version of such coupling is the use of quantum chemistry to calculate potentials for determining force fields in molecular dynamics simulations. For many electrochemical problems, this approach is inadequate and Car–Parrinello techniques are required. The interface between molecular dynamics and macroscopic theory is less developed. The largest effort has been directed to modeling of mechanical properties of solids (*14*). In the present volume an effort in this direction for electrochemical interfaces is reviewed in Chapter 2 by Walbran and Halley, in which the relevant macroscopic model is the Gouy–Chapman theory. We have also reported some preliminary attempts at rescaling of molecular dynamics for polymer systems elsewhere (*15*). In fact, what is needed is something like renormalization group theory (*16*) in statistical mechanics, which couples scales systematically. (Possibly it is more correct to say that we need a systematic, quantitative method for coarse graining molecular dynamics models, because no implication that the resulting coarse grained models will be self similar, as they are in the renormalization group theory of phase transitions, is suggested here.)

The subject of this volume is theory which is not purely phenomenological. However, as discussed above, to make contact between most experimental information and the most detailed possible calculations at the electronic scale, many orders of magnitude in space and time scale must be bridged. Thus we are necessarily dealing with a hierarchy of scales and methods. As a result, we find chapters that correspondingly range in method and scale from ab initio Hartree–Fock to classical molecular dynamics and beyond. However, in the present volume, I have chosen to group the contributions, not by method or scale, but by the class of physical and chemical system and phenomenon on which they report.

The chapters are arranged in an order in which the "easiest" problems are discussed first, with the increasingly challenging ones coming later in the volume. "Easiest" here means "most amenable to first principles treatment". We take these to be calculations of the structure of sp-metal-inert liquid interfaces. Tragically, David Lee Price, who has carried such calculations farthest and who participated in the symposium, was not able to contribute a chapter to this volume because he died suddenly in February 2000. I refer readers to his recent paper (17) on the cadmium–water interface. Next in difficulty are noble metal–liquid interfaces. Two chapters on copper–water interfaces illustrate the progress and the difficulties: in Chapter 1 Larry Curtiss' group illustrates that one can do cluster calculations to a high level of accuracy, but Chapter 2 by S. Walbran and Halley shows that though Car–Parrinello-like calculations can mimic copper capacitance, including the d-electrons of a noble metal–electrode in direct dynamics is still out of reach. Others who have carried out calculations on systems of this type include Kovalenko and Hirata (18). In this latter work (19), use is made of one of the several analytical theories of the liquid side of the electrode-electrolyte interface. (A difficulty with this approach as described in reference 18 is that the average solvent structure is used, whereas, within Born–Oppenheimer approximation, the electrons of the metal see the instantaneous solvent structure.) Chapter 3 from NMR experimentalist A. Wieckowkski's group shows one way to begin to extract detailed in situ experimental information about electronic structure at these interfaces.

In this collection, we present three chapters on theoretical studies of reactions at interfaces. In Chapter 4 Wolfgang Schmickler reviews the Anderson–Newns approach, in Chapter 5 M. Philpott's group presents a pioneering study of simulation of dissolution, and in Chapter 6, B. Smith presents innovative studies of wave packet dynamics studies of electron transfer at the semiconductor–liquid interface that reveal how much qualitative physics is buried or overlooked in traditional models. I regret that Gregory Voth, who contributed to the symposium, could not contribute to this book but refer readers to his group's work on electron transfer (20), which is similar to some of my own group's efforts (8), as well as to the Voth group's work on static structure of the interface (21).

The next group of five chapters refers to oxide–liquid (usually water) interfaces. These are immensely important in applications and also extremely difficult, not least because the solid-state physics of oxides is not well understood. In particular it is well known that the standard LDA approximations (*22*) cannot correctly predict from their composition whether bulk oxides will be insulators or metals. This is because the strong electron–electron correlations in oxides (*23*) make it possible for gaps, which lead to insulating behavior even though the corresponding noninteracting electron gas would be a metal, to appear in the electronic spectrum. (Disorder in the oxides at electrode–electrolyte interfaces can also lead to insulating behavior by another mechanism (*24*).) It is quite possible that as the potential drop across a metal–oxide–water interface changes, the oxide may, through one or more of these mechanisms, pass from metal to insulator, with significant effects on the electrochemical properties of the interface. No quantitative models of such behavior have yet been possible. The understanding of oxides is absolutely essential to the ultimate understanding and control of many types of metallic corrosion, which is limited by the stability and growth mechanisms of the passivating oxide layer on the metal surface. For that part of the problem, successful microscopic description at least of the structure and dynamics of point defects in oxides (*25*) will be required.

Even if one ignores the deficiencies of first principles LDA calculations for oxides cited earlier, formidable difficulties exist. Cutoffs for first principles electronic structure calculations using LDA plane wave methods are five or more times larger than those required for sp-metals, leading to calculational times for direct dynamics methods that are nearly two orders of magnitude greater and essentially out of reach presently. Further, water dissociation at these surfaces can seldom be neglected. The first chapter here, Chapter 7 by M. Ryan and co-workers, is experimental and describes recent work that elucidates the structure of the passivation layer on iron. This significant achievement gives modelers a starting point for studying passivation in iron for the first time. Chapter 8 by S. Parker and co-workers describes classical molecular dynamics calculations of oxide-water interfaces using shell model potentials. Chapter 9 by Rustad and Felmy describes calculations of hydroxylation of hematite using a dissociable, but classical model of water. Next, we have two chapters that include electronic structure in the description of the oxide: Chapter 10 by Truong and co-workers embed a cluster in a background of point charges. In Chapter 11, Schelling and Halley fit a self-consistent tight binding model to first principles results to describe the oxide-metal interface. (There is no water yet in the latter work though.)

Finally, we have two chapters describing complex liquid–solid problems that are not electrochemical. In Chapter 12, Yanhua Zhou et al. of the Goddard group presents results on sliding of one surface past another in the presence of a model lubricant using a realistic classical molecular dynamics model. In Chapter 13, Voue et al. present results on wetting. Whether further progress on these problems will really require more detailed treatment of electronic structure is less clear than in the other problems studied in this book.

The chapters in this volume describe an ongoing program. The goal of microscopic prediction is not reached for most of these systems. (sp-metal-water structure may be an exception.) In many cases, qualitative there are surprises with almost every new calculation, which should begin to convince doubters that simulation is more than just a matter of adding quantitative detail to the conclusions of analytical theory in this field. Eventually, we may find that some of the several computational methods used here dominate, but presently the diverse search for optimal computational methods is useful and healthy.

For all of these challenges, the talented group of participants represented in this symposium proceedings is making a good beginning, but clearly a larger effort could be justified. We hope that both funding agencies and scientists with related interests or who are searching for a new field of specialization will note this.

A. Wieckowski is thanked for inviting me to organize a symposium on this subject as a part of a series he is organizing for the American Chemical Society (ACS) Division of Colloid and Surface Chemistry on various aspects of electrochemistry and for guidance through the ACS bureaucracy during the project. I am grateful to several people who, though unable to be present at the ACS meeting in New Orleans, nevertheless agreed to contribute to these proceedings in order to achieve a more complete representation of the state of work of the field. The Petroleum Research Fund provided financial assistance for this symposium. Larry Curtiss is thanked for editorial assistance as are the several referees.

References

1. Faraday, M. *Phil.Trans.***1883**, 507–522; *Phil. Trans.***1833**, 675.
2. Gouy, G. *J. Phys. Radium* **1910**, *9*, 457; *Compt. Rend.* **1910**, *149*, 654; Chapman, D. L. *Phil. Mag.* **1913**, *25*, 475.
3. Marcus, R. *J. Chem. Phys.* **1956**, *24*, 966, 979; *J. Chem. Phys.* **1957**, *26*, 867, 872; *Can. J. Chem.* **1959**, *37*, 155; *Trans. Symp. Electrode Processes;* **1957**, page 239; *Trans. N. Y. Acad. Sci.***1957**, *19*, 423; *Dis. Fara. Soc.***1960**, *29*, 21; *J. Phys. Chem.* **1963**, *67*, 853, 2889; *J. Chem. Phys.* **1965**, *43*, 679.
4. von Helmholz, H. L. F. *Ann. Phys.* **1853**, *89*, 211; *Ann. Phys.* **1879**, *7*, 337.
5. Halley, J. W.; Walbran, S.; Price, D. L. *Adv. Chem. Phys* **2000**, *116*, 337
6. Car, R.; Parrinello, M. *Phys. Rev Lett.* **1985**, *55*, 2471.
7. Halley, J.W.; Walbran, S.; Price, D. L. in *Interfacial Electrochemistry: Theory, Experiment, and Applications*; A.Wieckowksi, ed.; Marcel Dekker: NY, 1999; pp 1-17.
8. Halley, J. W.; Smith, B. B.; Walbran, S.; Curtiss, L. A.; Rigney, R. O.; Sujiatano, A.; Hung, N. C.; Yonco, R. M.; Nagy, Z. *J. Chem. Phys.* **1999**, *110*, 6538.

9. Family F. in *Electrochemical Synthesis and Modification of Materials. Symposium.* Materials Research Society: Pittsburgh, PA, 1997; pp 123–140.
10. Daikin, L. I.; Urbach, M. *Phys. Rev.* **1999,** *E59,* 1821.
11. Kolb, D. M. *Prog. Surf. Sci.* **1996,** *51,* 109.
12. Chatterjee, P. K; Doering, R. R. *Proc. IEEE* **1998,** *86,* 176.
13. Li, X-P; Nunes, R. W.; Vanderbilt, D. *Phys. Rev.* **1994,** *B47,* 10891; Nunes, R. W.; Vanderbilt, D. *Phys.Rev.* **1994,** *B50,* 17611.
14. *Multiscale Modelling of Materials, Symposium of Materials Research Society;* Bulatov, V. V; de la Rubia, T.D.; Phillips, R.; Kaxiras, E.; Ghoniem, N., eds., Materials Research Society: Pittsburgh, PA, 1999.
15. Halley, J. W.; Schelling, P.; Duan, Y. *Electrochim. Acta,* in press.
16. Fisher, M. E. *Rev. Mod. Phys.* **1998,** *70,* 653.
17. Price, D. L. *J. Chem. Phys.* **2000,** *112,* 2973.
18. Kovalenko, A.; Hirata, F. *J. Chem. Phys.* **1999,** *110,* 10095.
19. Rickayzen, G. in *Microscopic Models of Electrode–Electrolyte Interfaces*; Halley, J. W.; Blum, L., eds; Electrochemical Society: Pennington, NJ, 1992; pp 93-102,143; Boda, D.; Henderson, D.; Chan, K-Y.; Wasan, D.T. *Chem. Phys. Lett* **1999,** *308,* 473; Kinoshita, M.; Hirata F. *J. Chem. Phys.* **1996,** *104,* 8807.
20. Straus, J. B.; Calhoun, A.; Voth, G. A. *J. Chem. Phys.* **1995,** *102,* 529.
21. Straus, J. B.; Voth, G. A. *J. Chem. Phys.* **1993,** *97,* 7388.
22. *Local Density Approximations in Quantum Chemistry and Solid State Physics. Proceedings of a Symposium on Local Density Approximations in Quantum Chemistry and Solid State Theory;* Dahl, J. P.; Avery, J., eds. Plenum: New York, **1984.**
23. Tokura,Y.; Nagaosa, N. *Science (Washington, DC)* **2000,** *288,* 462.
24. *Proceedings of International Conference. Localization 1999: Disorder and Interaction in Transport Phenomena;* Hamburg, Germany, 1999; *Annalen Phys.* **1999,** *8 (7-9).*
25. Chao, C. Y.; Lin, L.F.; MacDonald, D. D *J. Electrochem. Soc.***1981,** *128,* 1187; *J. Electrochem. Soc.* **1981,** *128,* 1194; *J. Electrochem. Soc.* **1982,** *129,* 1874.

J. WOODS HALLEY
School of Physics and Astronomy
University of Minnesota
Minneapolis, MN 55455

Electronic Properties of the Metal–Solvent Interface

Chapter 1

Electronic Structure Studies of the Interaction of Water with a Cu(100) Surface

Peter Zapol[1], Conrad A. Naleway[1], Peter W. Deutsch[2], and Larry A. Curtiss[1]

[1]Materials Science and Chemistry Divisions, Argonne National Laboratory, 9700 South Cass Avenue, Argonne, IL 60439
[2]Department of Physics, Pennsylvania State University, Monaca, PA 15061

The results of a density functional study of the chemisorption of water on a Cu(100) surface are presented. Both atomic cluster and periodic supercell models of the surface were used in the investigation. From the cluster studies a single water molecule is bound by about 0.6 eV to the surface and is in an on-top site. The addition of a second water molecule in a site adjacent to the first one is not favorable due to polarization of the electron density near the surface. The periodic density functional calculations give results consistent with the cluster studies.

The nature of the interaction of water with noble metal surfaces is of great importance in electrochemistry, corrosion, and heterogeneous catalysis. There is evidence that intermolecular interactions between water molecules can compete with water-surface interactions, although little is known about the role that hydrogen bonding between water molecules plays in determining the structure of the metal/water interface.

Very little is known experimentally about the nature of the interaction of a single water molecule with noble metal surfaces because it is difficult to separate out the water/water effects from the water/surface interactions. In a study of water on a Au(111) surface, Kay et al.(*1*) have reported that the temperature programmed desorption (TPD) spectrum does not display a well-resolved submonolayer peak indicating that H_2O binds more strongly to itself than to the Au substrate. Arrhenius analysis of the TPD peak gave a binding energy of about 0.4 eV. In an electron-energy-loss spectroscopic (EELS) study of water on a Cu(100) surface , Andersson et al. (*2*) found that it adsorbs with the oxygen end towards the surface and its molecular axis significantly tilted relative to the surface normal.

4

There have been several theoretical studies of a single water molecule adsorbed on Cu and Ag surfaces with the surface represented by atomic clusters. Kuznetsov and Reinhold (*3*) studied an H_2O molecule adsorbed on Cu_5 and Ag_5 clusters with the CNDO/2 semiempirical molecular orbital method. They found the most stable sites to be hollow sites (4-fold coordination) with binding energies of 1.5 and 1.1 eV, respectively, which they indicated were overestimates. Ribarsky et al.(*4*) reported local density $X\alpha$ calculations of bonding between H_2O and a Cu_5 cluster. They found a binding energy of 0.38 eV for water positioned at the on-top site with a tilt angle of about 70° from the normal. Rodriquez and Campbell (*5*) studied H_2O on a Cu surface using the INDO/S method with the surface represented by 14-18 atom clusters to determine the nature of the bonding. Ignaczak and Gomes (*6*) reported a density functional study of the interaction of H_2O with cluster models of copper, silver and gold surfaces. The clusters contained up to 12 atoms. A minimal basis set with effective core potentials was used. They found that on-top and 2-fold sites had similar binding energies of about 0.3 eV on the Cu_{12} clusters with the water molecules tilted in both cases.

In this paper we report all-electron gradient-corrected density functional calculations of the chemisorption of water on a copper surface to determine the most favorable adsorption site, binding energy, and the orientation of the water molecule. The calculations are done on clusters of Cu atoms and a supercell model of the surface.

Theoretical Methods

The H_2O/Cu surface calculations were carried out on clusters chosen to model the Cu(100) surface. Two cluster sizes were used in most of the calculations: (1) a ten atom cluster containing five atoms in the first layer, four in the second, and one in the third and (2) and fourteen atom cluster containing eight atoms in the first layer and six in the second. These clusters are illustrated in Figure 1. The experimental crystal structure was used to construct the clusters (the Cu-Cu interatomic distances in the layers is 2.54 Å). The water molecule was held rigid at its experimental geometry: $R(OH) = 0.957$ Å and $<HOH = 104.5°$. The cluster calculations were done with a 9s5p3d contraction of Wachter's (*7*) 14s9p5d primitive set for Cu and an f polarization function. This is referred to as "6-311G*" in the Gaussian94 code (*8*). The standard 6-311G* basis set was used for oxygen and hydrogen in the calculations. Most of the calculations were done at the B3LYP level of theory. In addition, Hartree-Fock (HF) and second order perturbation theory (MP2) calculations were done for comparison with the B3LYP results.

Periodic density functional calculations reported in this paper utilized crystalline orbitals constructed from the localized atomic orbitals, which are linear combinations of Gaussian orbitals. The Kohn-Sham equations for crystalline orbitals were solved self-consistently using the nonlocal exchange functional of Becke (*9*) and the nonlocal correlation functional of Lee, Yang and Parr (*10*) (BLYP). In the periodic calculations the 6-31G* (*11*) basis set for copper was slightly truncated and reoptimized for crystal studies: the outer sp-function and f-function were removed from the basis set and the 3sp contraction was split into 2 and 1; the latter was optimized in the bulk copper crystal to 0.12664. The standard 6-31G* basis set was used for oxygen and hydrogen. The optimized copper crystal lattice constant of 3.53 Å was obtained with this basis set. This agrees reasonably with the experimental value of 3.59 Å. The surface calculations were done using a periodic slab model with 21 k-points in the planar Brillouin Zone. The periodic calculations were done with the CRYSTAL95 (*12*) code.

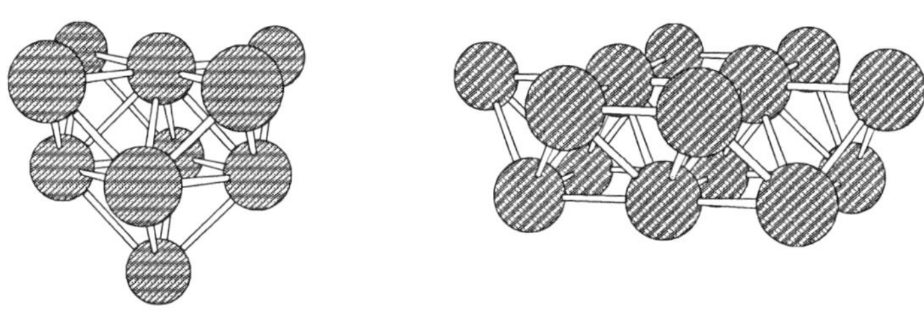

Figure 1. *Copper atom cluster models of (100) surface. On the left is a ten atom cluster and on the right is a fourteen atom cluster.*

Results and Discussion

Cluster Models. Initially, the dependence of the chemisorption energy on the level of the theory was investigated for a single water molecule adsorbed on the on-top site of the Cu_{10} cluster. In addition, the distance of the water molecule to the surface was optimized. The results are shown in Table I. At the HF level the water molecule is bound by 0.37 eV with an optimized water-surface distance of 2.27 Å. Inclusion of correlation at the B3LYP increases the binding to 0.60 eV and shortens the distance to 2.10 Å. The BLYP density functional gives a similar result as the B3LYP functional (0.63 eV and 2.10 A). The inclusion of correlation by second order perturbation theory (MP2) gives a binding energy of 0.72 eV, which is consistent with the B3LYP results. Thus, the B3LYP/6-311G* method was used in all of the other cluster calculations reported here.

The dependence of the adsorption energy on cluster size was investigated using four different cluster sizes: the ten and fourteen atom clusters described in theory section as well as a nine atom cluster (formed by removing the one atom layer from Cu_{10}) and a sixteen atom cluster (two eight atom layers). The B3LYP/6-311G* results for on-top adsorption for the four different clusters are given in Table II. With the exception of the Cu_9 cluster that has a binding of 0.36 eV, the other three cluster sizes (Cu_{10}, Cu_{14}, Cu_{16}) have binding energies that are reasonably similar: around 0.6 – 0.7 eV. The Cu_9 binding may be small because the cluster has open shell electronic configuration. Table II also contains the results for adsorption at different sites for the Cu_{14} cluster. The most favorable site is the on-top site with a binding of 0.72 eV, the 2-fold and 4-fold sites have binding energies of 0.40 and 0.32 eV, respectively. The interaction of the water molecule with the hydrogens pointed down is attractive and has a binding of 0.43 eV.

Table I. Comparison of different methods for H_2O interaction with the on-top site of the Cu_{10} cluster

Method	Basis Set[a]	R,[b] Å	ΔE, eV
HF	6-311G*	2.27	0.37
BLYP	6-311G*	2.06	0.63
B3LYP	6-311G*	2.10	0.60
MP2	6-311G*	$(2.10)^c$	0.72

[a] Basis set defined in theoretical methods section.
[b] Distance for oxygen of H_2O and the closest copper atom of the Cu_{10} cluster.
[c] Held fixed at the B3LYP level of theory.

Table II. B3LYP/6-311G* results for H_2O adsorption on Cu clusters

Adsorption site	Cluster	R,[a] Å	ΔE, eV
1-fold	Cu_9 (5,4)	2.1	0.36
	Cu_{10} (5,4,1)	2.1	0.60
	Cu_{14} (8,6)	2.0	0.72
	Cu_{18} (8,8)	$(2.1)^b$	0.60
2-fold	Cu_{14} (8,6)	1.9	0.40
4-fold	Cu_{14} (8,6)	1.9	0.32
1-fold (H-down)	Cu_{14} (8,6)	2.9	0.43

[a] Distance for oxygen of H_2O and the closest copper atom of the Cu_n cluster.
[b] Held fixed at the Cu_{10} result.

We also investigated the tilt of the water molecule and found that at the B3LYP/6-311G* level the water molecule on the 1-fold site tilts by about 30° relative to the surface normal. The energy lowering gained by this tilt is only 0.03 eV.

The adsorption of a second molecule at a site adjacent to the first water in an on-top site was investigated using the B3LYP/6-311G* method. The structure of the two water molecules on the surface of the Cu_{14} cluster is shown in Figure 2(a). Both molecules were held fixed at 2.10 Å from the surface. The binding energy of the second water molecule is only 0.03 eV and, if allowed, the second water molecule moves away from the surface. The adsorption of a hydrogen bonded water dimer on the surface was investigated. The structure for the water dimer is shown in Figure 2(b). In this case the second water molecule has an adsorption energy of 0.38 eV. The increase over the adjacent adsorption site is due to the formation of the hydrogen bond with the first water molecule.

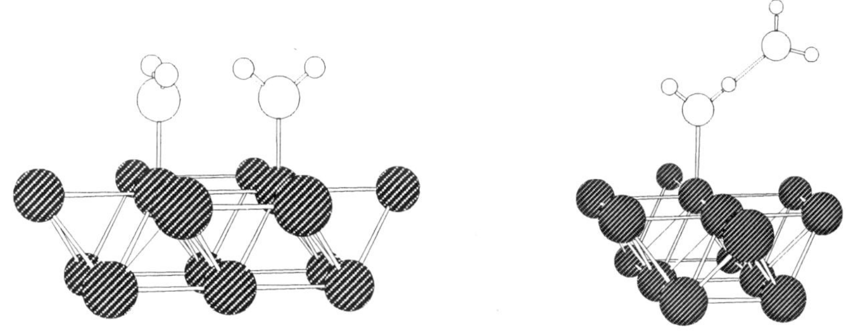

Figure 2. *(a) Orientation of two waters in adjacent sites on cu14 cluster and (b) hydrogen bonded water dimer on the surface*

The unfavorable interaction of the second water molecule in an adjacent site on the Cu(100) surface can be understood in terms of the charge redistribution that occurs near the surface. The Mulliken charges on the surface copper atoms of the Cu_{14} cluster are shown in Figure 3. They indicate that the adsorption of a water molecule in an on-top site causes polarization of the charge on the surface atoms so that the nearest neighbor atoms of the Cu atom to which the water forms a bond are negatively charged. This results in an unfavorable interaction of the second water molecule in the adjacent site. The repulsion of the two waters will also contribute to the unfavorable interaction energy for the adjacent site. This cluster result is consistent with the work of Price and Halley (*13*) who reported on a simulation of the water/copper interface using a molecular dynamics density functional plane wave approach. From these simulations they found that polarization of the electron density near the surface resulted in only a small number of the waters close to the surface.

Periodic Calculations. A two-dimensional periodic copper slab consisting of two layers was used for surface modeling. The lattice constant was fixed at the experimental value for consistency with cluster calculations. The water molecules were adsorbed at 50% of the surface copper atoms, thus giving half-monolayer coverage of the surface as shown in Figure 4. The orientation of all water molecules is the same and corresponds to the most stable geometry found in cluster calculations with the oxygen end of the molecule directed toward surface. The distance between the molecule and the surface was optimized while their geometries were kept frozen. An equilibrium distance of 2.14 Å was obtained, similar to that found in the cluster calculations. The binding energy of water to the copper surface was found to be 0.27 eV at this coverage. The water molecules at half-monolayer coverage repel each other due to dipole-dipole interactions. To extrapolate water-surface binding energy to a low coverage limit we calculated the water repulsion energy separately and added it to the binding energy. The corrected binding energy is 0.33 eV.

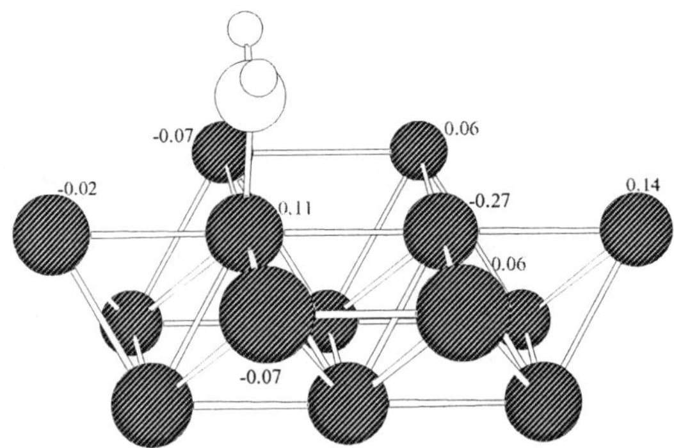

Figure 3. *Mulliken charges on the top layer of the Cu$_{14}$ cluster with H$_2$O adsorbed at an on-top site.*

It was found experimentally that water molecules tilt when adsorbed on the (100) copper surface. If we consider this tilt at half of the monolayer coverage, the water-water repulsive interaction is greatly reduced when molecules are tilted and water molecules start to attract each other at tilt angle of about 35°. Tilt is estimated to produce a very modest gain of about 0.03 eV in the water-surface binding energy. Since the energy contribution of water-water interactions is much higher than that extrapolation of tilt angle to low coverage limit is not possible.

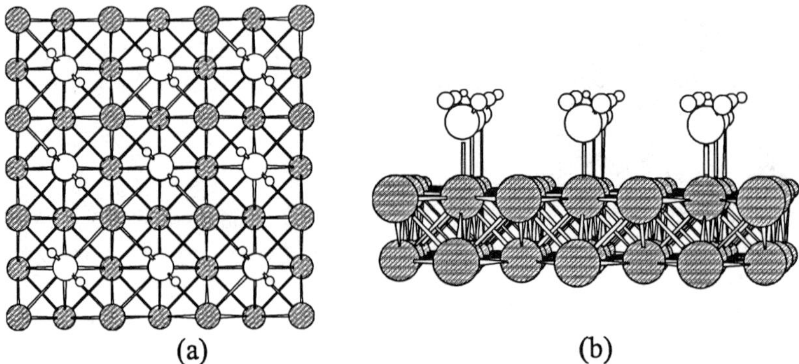

(a)   (b)

Figure 4. *Top (a) and side (b) views of the periodic two-layer slab model for (100) copper surface with adsorbed water at half of the monolayer coverage. Large dark spheres are copper atoms in the first layer and small dark spheres are in the second layer. Hydrogen and oxygen appear as small and large white spheres, respectively.*

Conclusions

The results of a density functional study of the chemisorption of water on a Cu(100) surface are presented. Both atomic cluster and periodic supercell models of the surface were used in the investigation. From the cluster studies a single water molecule is bound by about 0.6 eV to the surface and in an on-top site. The addition of a second water molecule in a site adjacent to the first one is not favorable due to polarization of the electron density near the surface. The periodic density functional calculations give similar results for the geometries and energies of the H_2O/Cu adsorption.

Acknowledgements

We gratefully acknowledge use of the advanced computing resources at Argonne's Center for Computational Science and Technology. This work is supported by the U.S. Department of Energy, BES-Materials Sciences, under Contract W-31-109-ENG-38.

Literature Cited

1. Kay, B. D.; Lykke, K. R.; Creighton, J. R.; Ward, S. J. *J. Chem. Phys.* **1989** *91*, 5120.
2. Andersson, S.; Nyberg, C.; Tengstal, C. G. *Chem. Phys. Lett.* **1984** *104*, 306.
3. Kuznetsov, A.; Reinhold, J. *Z. Phys. Chemie, Leipzig* **1986** *267*, 824.
4. Ribarsky, M. W., Luedtke, W. D.; Landman, U. *Phys. Rev. B* **1985** *32*, 1430.
5. Rodriguez, J. A.; Campbell, C. T. *Surf. Sci.* **1988** *197*, 567.
6. Ignaczak, A.; Gomes, J. A. N. F. *J. Elect. Chem.* **1997** *420*, 209.
7. Wachters, A. J. H. *J. Chem. Phys.* **1970** 52 (1970).
8. *Gaussian 94*, Frisch M. J. et al, Gaussian, Inc., Pittsburgh PA 1996.
9. Becke, A. D. *Phys. Rev. A* **1988** *38*, 3098.
10. Lee, C.; Yang, W.; Parr, R. G. *Phys. Rev. B* **1988** *37*, 785.
11. Rassolov, V.; Pople, J. A.; Ratner, M. A.; Windus, T. L. *J. Chem. Phys.* **1998** *109*, 1223.
12. Dovesi, R.; Saunders, V. R.; Roetti, C.; Causa, M.; Harrison, N. M.; Orlando, R.; Apra, E. *CRYSTAL95 User's Manual*, University of Torino, Torino, 1996.
13. Price, D. L.; Halley, J. W. *J. Chem. Phys.* **1995** *102*, 6603.

Chapter 2

Direct Dynamics Simulations of the Copper-Water Interface: Successes and Problems

Sean Walbran[1] and J. W. Halley[2]

[1]Forschungszentrum Jülich GmbH, Institut für Werkstoffe und Verfahren der Energietechnik (IVW–3), D–52425 Jülich, Germany
[2]School of Physics and Astronomy, University of Minnesota, Minneapolis, MN 55455

We report on direct dynamics studies of the electrode/electrolyte interface, in which the electronic structure of the electrode is calculated at each time step of a molecular dynamics simulation. Results for a model of the copper/water interface, in which copper pseudopotential centers are described with 4s electrons, are presented for two crystal faces. We discuss of our attempts to model the 3d electrons on the electrode using a number of new approaches, and describe the computational challenges for such a system. Finally, we describe recent work applying direct dynamics techniques to the aluminum/water interface.

Introduction

Processes and reactions at the electrode/electrolyte interface are fundamental to such technologies as batteries, fuel cells, photochemical cells, and chemical etching. They are at the root of electrochemistry, and play a large role in corrosion, stress corrosion cracking, toxic chemical abatement, and weathering processes important to industry and geoscience. The interplay among the solid electrode, liquid solvent, and ions of the electrolyte presents a significant challenge to traditional theoretical modeling, involving physics on a wide range of length and time scales. We have used direct dynamics simulations to study the electrode/ electrolyte interface, including the electronic structure of the electrode in a molecular dynamics simulation of the aqueous electrolyte. Here, we review our calculations(1) of the copper/water interface and its capacitance, discuss some attempts to extend the simulations to include the d electrons on copper, and report on recent results with an aluminum electrode.

The Copper–Water Interface

The electrostatic potential drop across the electrode–electrolyte interface as a function of charge on the electrode is a fundamental property of the interface. It is known that the measured capacitance of metal/water interfaces includes contributions from the electrons of the metal electrode (2– 9) , from the water near the surface ('Stern layer' contributions), and from the ions of the electrolyte. These contributions are not neatly separable, though they have sometimes been modeled as three capacitors in series. Calculations carried out in which the electronic and 'Stern layer' contributions are modeled together using a direct dynamics model for the interface(*10*) are described here. We model the ionic contributions using the traditional Gouy Chapman theory (*11*) but employing a method for determining the boundary condition for the continuum Gouy Chapman theory from the microscopic direct dynamics model. The results give a relatively parameter free representation of the electrostatic response.

As discussed in references 2– 9 and elsewhere, the response of the electrons of the metal to changes in the electrostatic potential of the metal relative to the electrolyte results in a significant contribution to the differential capacitance. It is also well known that the dielectric properties of the water near the metal–electrolyte interface contribute to the capacitance, as first recognized by Stern (*13*) . Both the electronic and 'Stern Layer' effects are treated together without artificial spatial separation in a first principles model (*10*) in which wave functions are recalculated together with the corresponding forces on the water molecules in the model at each step of a molecular dynamics simulation. The model used for the calculations described here is identical to that described in reference 10 except that an improved central potential model (*14*) has been used to model the water. We also modified the code to model the 111 as well as the 100 surfaces of copper. The model has the virtues of the one in reference 10 but also its limitations. In particular, only the s–like electrons of the metal are treated explicitly.

We studied the charge on the electrode as a function of the potential drop across the interface. Because we found that equilibration of the water in response to a change in the field took a long time (of order 50ps) it was found expedient to run a classical simulation (*14*) at each new field before initiating the direct dynamics equilibration at each field. Explicitly the simulation sample consisted, for the 100 surface, of 245 water molecules and 180 copper atoms periodically continued, in a cell of dimensions 80 au x 29 au x 29 au as in reference 9. As in that paper, for the 100 surface, the electronic wavefunctions were calculated in a 'small cell' of dimensions 40 au x 29 au x 29 au . For the 111 surface the water sample contained 216 water molecules, the unit cell had the dimensions 61.4 au x 33.6 au x 33.6 au (32.5 Å x 17.8 Å x 17.8Å), and the 'small cell' had dimensions 30.7 au x 33.6 au x 33.6 au . The time step used for integrating the motion was 0.5 fs. The sample was allowed to relax for approximately 80,000 time steps in the classical simulation, followed by 2,000 time steps of equilibration in the quantum mechanical simulation with the electrons frozen in place. Subsequently the system was run with full coupling between the electrons and all atomic centers for a further 3,000 time steps, with statistical measurements made during the final 1,000 time steps. The sample was periodically repeated an infinite number of times in all three dimensions as described in reference 10 .

Though we can simulate the Stern and electronic contributions quite realistically with this model, a realistic simulation of the double layer region, including the ions of the electrolyte explicitly, remains beyond our simulation capacity at present if we wish to include a microscopic description of the solvent and electronic structure of the electrode in the simulation. On the other hand, because the length scale variations of the potentials in the double layer region are also larger, at low (~.001 – .01 M) ionic strengths, by about an order of magnitude, the mean field representation given by the classical Gouy–Chapman theory (*11*) is rather good in this region. To determine the capacitance of the entire system, we have coupled this theory in a self–consistent manner to the model for the region near the electrode described above.

One approach is to simply add the Gouy–Chapman capacitance in series with the capacitance of the Stern layer. This approach implicitly and unrealistically postulates a sharp interface between the Stern layer and the region in which the electrolyte can be treated by the Gouy–Chapman model. If such a sharp interface existed, then the boundary condition on the macroscopic electric field in the Gouy Chapman theory would be simply 4π times the charge σ applied to the electrode, leading to the usual series expression for the lumped capacitances. However, when the interface is not sharp, the boundary condition for the Gouy Chapman model can be estimated from the molecular dynamics simulation by calculating the macroscopic electric field at the plane $z = zc$ at which the model begins to represent the system by the Gouy–Chapman model. (This can be shown to be equivalent in the linear response regime to taking account of the nonlocal and/or inhomogeneous character of the dielectric response of the solvent.) The appropriate value of z_c is not known a priori.

Because the $E(z_c)$ so determined is not necessarily equal to $4\pi\sigma/\varepsilon_{bulk}$ where ε_{bulk} is the bulk dielectric constant of the solvent, then we cannot simultaneously 1) use $E(z_c)$ as a boundary condition, 2) retain charge neutrality of the interface, and 3) use the Gouy Chapman theory in its usual form with a constant value of ε for the solvent. A simple and physically reasonable path out of this situation is to adopt an extension of the Gouy Chapman model in which ffl varies smoothly with z from the simulation determined value $\varepsilon(z_c) = 4\pi\sigma/E(z_c)$ to the bulk value far from the electrode:

$$\varepsilon(z) = \varepsilon(z_c) + \left(\varepsilon_{bulk} - \varepsilon(z_c)\right)\left(1 - e^{\frac{-(z-z_c)}{z_o}}\right)$$

We found that the net capacitance which we calculated from the resulting model depended only weakly on the decay length, which was taken to be $z_o = 5\text{Å}$. (Taking ε

$z_o = 5 \pm 2\text{Å}$ results in potential variances of $< 0.02V$ in capacitance plots such as those shown below.) With this form for $\varepsilon(z)$ we numerically solved the Poisson–Boltzmann equation for $z > z_c$, yielding the net potential drop from inside the electrode to the bulk of the solution as

$$\phi = \left(\phi_{MD}\left(z \to -\infty\right) - \phi_{MD}\left(z = z_c\right)\right) + \left(\phi_{GC}\left(z = z_c\right) - \phi_{GC}\left(z \to \infty\right)\right)$$

where $\phi_{MD}(z)$ is the electrostatic potential obtained from the simulation described in the last section. In practice, $z \to \infty$ is replaced by a point inside the model metal slab where the macroscopic field is essentially zero.

To obtain this $E(z_c)$, we calculated smoothed values of the macroscopic electric field E as found from the simulations at various planes z as a function of the electrode charge σ. The smoothing was accomplished by averaging the calculated fields over the xy plane and around each z value with a Gaussian weight of width 5 Å for the 111 surface and 7 Å for the 100 surface; this removes irrelevant fluctuations due to detailed molecular structure and corresponds to the standard definition of the macroscopic electric field (*17*). (The appropriate width of the Gaussian is not defined a priori and was chosen to be as small as possible to retain information about the change in field with distance from the electrode while minimizing noise; we were able to use a smaller width for the 111 surface in part because the simulation's cross–sectional area gave more water molecules per unit distance and hence less noise in the raw simulation data.)

An interesting feature of these results is that the macroscopic field can be lower near the interface than it is deep in the solvent. This is qualitatively different from results obtained from classical models (*15*), in which the fields near the interface are almost always higher than they are in bulk. The lower fields near the surface are due to the screening of the fields by the electrons of the metal, giving a smooth rise of the field from the bulk metal to the bulk solvent. We illustrate this effect explicitly in Figure 1.

Using the calculated macroscopic fields as boundary conditions, we determined the total electrostatic response of the interface with results shown for the two faces in Figures 2, 3, and 4.

Comparison with Experiment

The only undetermined parameter in this model for the structure of the interface is z_c, the point at which the continuum and simulation models are matched. In Figure 2 we show the comparison with experimental data from reference 17 for the 100

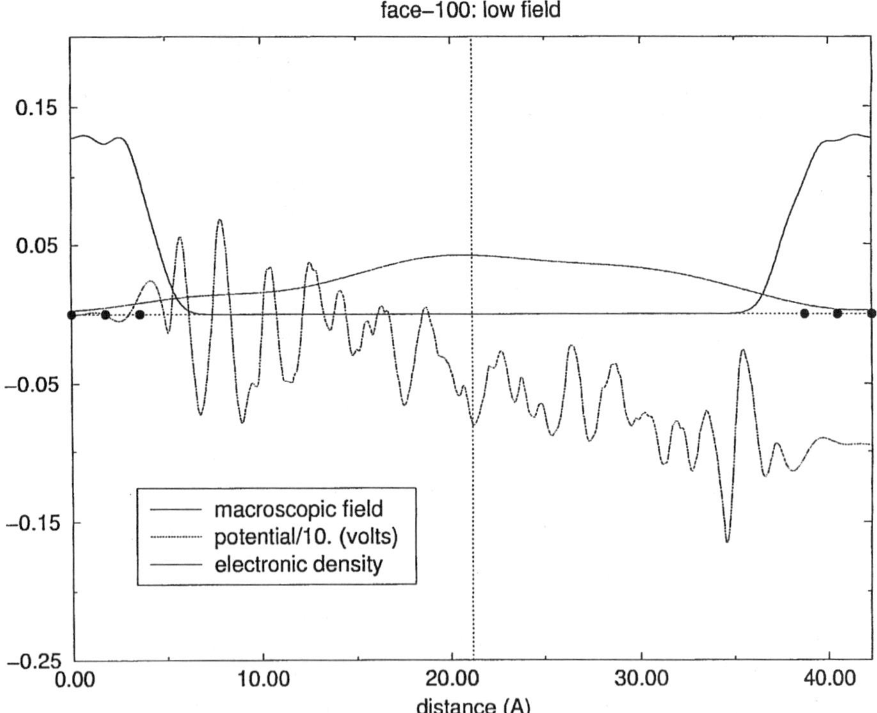

Figure 1. Electrostatic potential and clculated macroscopic electric fields are shown as a function of z for a fixed value of the charge σ on the electrodes. For reference purposes, the electron density is plotted along with solid circles indicating the position of copper centers. Units of the quantities shown are chosen for presentation purposes as E in V/Å , φ in V divided by 10 and electon density in e/au³ multiplied by 10. The 100 face is shown at charge σ= 10.3. Reprinted by permission from reference 1.

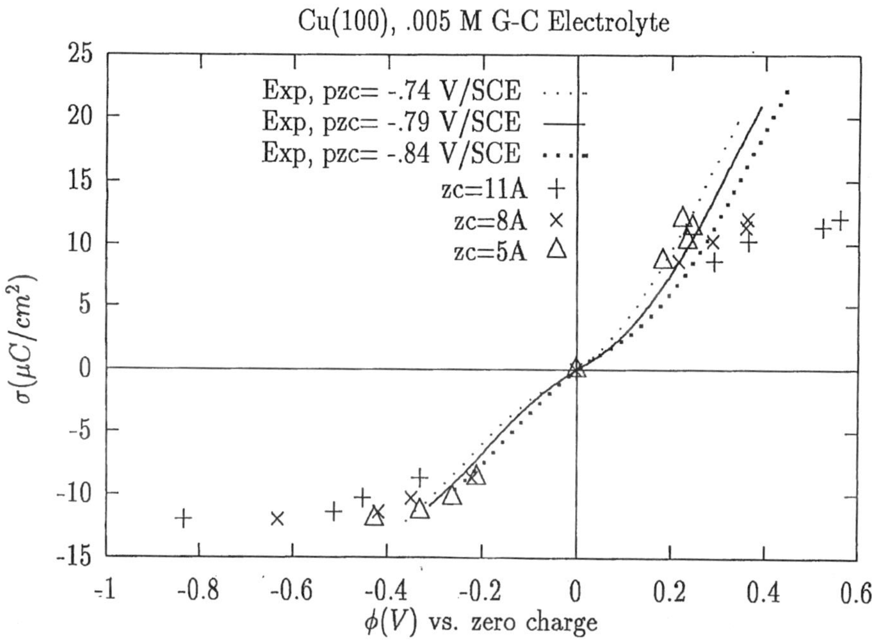

Figure 2. Calculated charge versus total potential drop for a 0.005 molar 1–1 electrolyte and various values of the matching distance z_c for the 100 surface. Integrated experimental results are shown from reference 17 with the zero of potential for the experimental results indicated. The reported potential of zero charge of reference 17 is −0.79 V/SCE for this surface. Reprinted by permission from reference 1.

surface, and in Figures 3 and 4 we compare experimental data from references 18 and 19 for the 111 surface. In each case we have shown the experimental results integrated to give $\sigma(\phi)$ and have chosen the zero of potential as the reported potential of zero charge of reference 18, namely −0.45 V/SCE and −0.79 V/SCE for the 111 and 100 surfaces, respectively. In addition we show the experimental curves assuming a potential of zero charge displaced by ±0.05V from the above values. For each surface the magnitude of the calculated capacitance is roughly consistent with the reported numbers of reference 18. For the 111 face, our capacitance values are larger than the results of reference 19 for all values of z_c. With regard to potential of zero charge for the copper–water interface, our results are not in disagreement with those of reference 17 . The experimental results are subject to significant uncertainties due to the reactivity of the copper electrode (*18– 21*), including surface oxidation, of which we take no account in the model.

The model of the electrostatic response of a copper electrode presented here exhibits features which are unique to our direct dynamics approach including the electronic structure of the electrode. The electron density at the surface displays an asymmetric response to electrode charging, as predicted qualitatively by more primitive models, but is here coupled correctly to the dynamics of the solvent and the Gouy–Chapman ionic screening. The macroscopic electric fields near the electrode–solvent interface are much more strongly screened than they are in models in which the electrode is treated classically.

Attempts at copper with d electrons

We examined the possibility of using direct dynamics simulation techniques like those used in the studies described above for explicit calculation of the equilibrium energetics and barrier heights in cuprous–cupric electron transfer, as we have done with classical models (*12*), by inclusion of a dissolved copper ion in the simulation. Earlier work done with the direct dynamics code has focused on bare electrodes in water, using purely local, semi–empirical pseudopotentials with which only the structure of the 4s electrons of the copper were calculated. Describing electron transfer with ionized copper centers requires a much larger calculation than these earlier efforts. Because the electronic structure of neutral copper is $[Ar]3d^{10}4s^1$, it is clear that a description of the oxidation process demands a simulation in which the 3d electrons are described explicitly. Tracking the dynamics of these more tightly bound electrons implies an increase in the number of wavefunctions calculated in the simulation as well as a substantial increase in the plane–wave cutoff energy (equivalently, a finer real–space mesh).

Table 1. Computational Cost of the Algorithm

Routine	*Scaling*	*Share (%, approximate)*
Orthogonalization	$M_\psi^2 N_G$	45
$H\psi$	$M_\psi N_G$	30
FFT	$M_\psi N_G \log N_G$	20

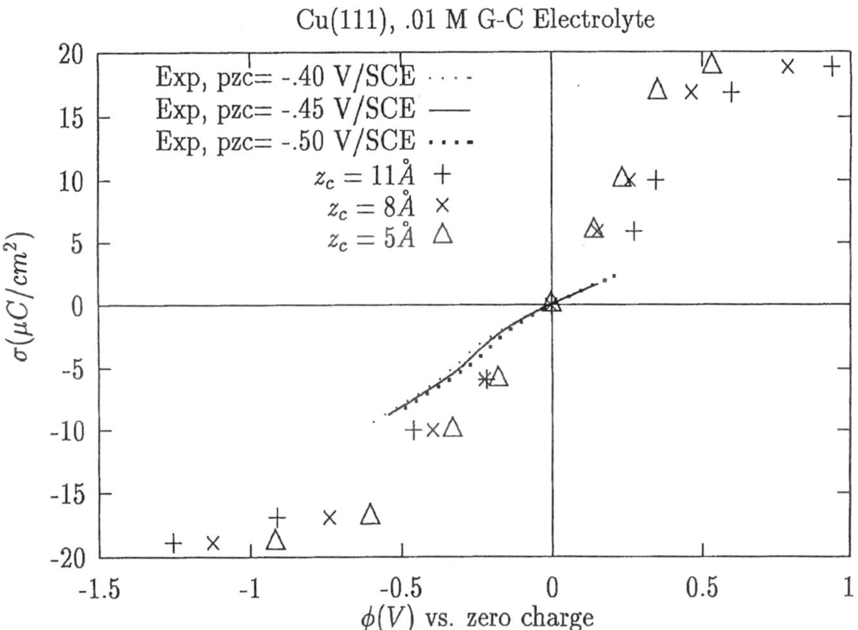

Figure 3. Same as Figure 2 , but for the 111 surface of copper. The reported potential of zero charge of reference 17 is −0.45 V/SCE for this surface. Reprinted by permission from reference 1.

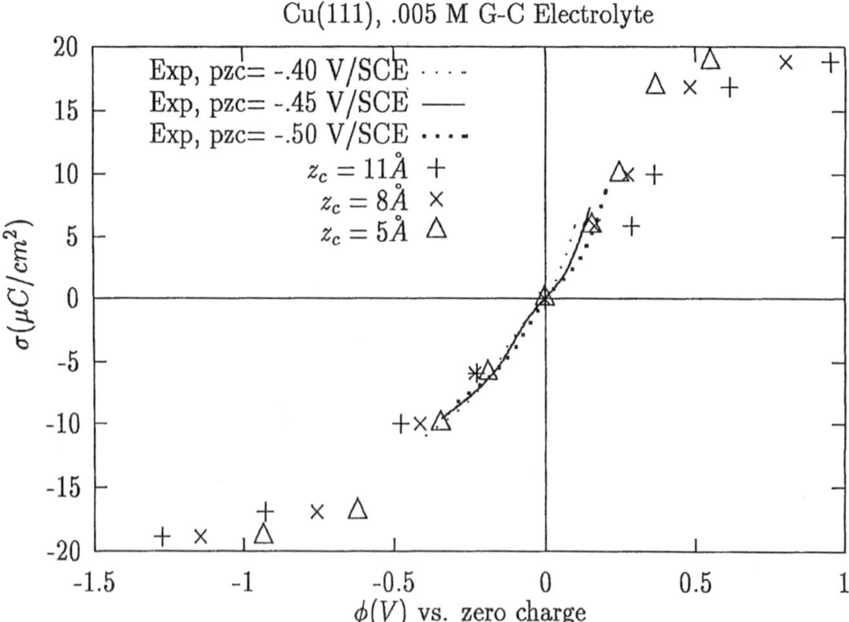

Figure 4. Same as Figure 3, but compared with the data of reference 18 , both theory and experiment with electrolyte concentration 0.01 M.Reprinted by permission from reference 1.

In the direct dynamics program, the computational cost of running the code on a typical system is dominated by three routines, as shown in table 1; the share of these routines of the total cost will vary depending on the values of M_ψ, the number of wavefunctions, and N_G, the number of plane–waves/grid points. The fourier transform ("FFT") is used in evaluating Hψ as $H\psi = T_G\psi_G + F^{-1}\left(V_R F\psi_G\right)$, where subscripts G and R refer to reciprocal and real space, respectively, T and V are the kinetic and potential energy operators, and F is the fourier transform operator.

To evaluate the LDA code for possible use in the electron transfer study, we carried out a number of preliminary calculations. The first was a calculation of the eigenvalues of a single, neutral 3d4s–copper atom (in vacuum with periodic boundary conditions). In agreement with the work from which the pseudopotential was derived(22), we found that a cutoff energy of about 80 Ry was necessary to converge the eigenvalues to satisfactorily match those of the all–electron calculation which generated the pseudopotential. This is a quite substantial increase over previous calculations, which were limited to 5 – 15 Ry. Calculations of an ionized atom yield energy differences which are in good agreement (error less than 3 percent) with experimental ionization energies, though absolute error values are significant on the electrochemical scale of tenths of volts.

To perform simulations of the electrode/electrolye interface, a simulation cell with a side length of 10 to 20 Å is required in order to minimize effects of the sample boundary on the system (in particular, the liquid component). For a copper electrode, this implies a slab with 5 to 6 atoms on a side, for a total of roughly 150 copper centers in the sample. The computational consequences of explicitly calculating the electronic structure of the 3d electrons on every copper atom are extreme; assuming the computation time scales with the orthogonalization step, which as described above is the most expensive part of the calculation, the increase in computation time for a fully described copper slab is on the order of 1000, which is prohibitively expensive at this time. To reduce this cost to the level of feasibility, several modifications were considered, including modification of the bases used to expand the wavefunctions, treating the electrode as a mixture of the 4s–copper in the bulk and the 3d4s–copper in regions where greater precision is necessary, and new methods of obtaining the ground state electronic density.

Initially, we considered using a more localized basis set than the traditional plane wave basis, inspired by the thought that the 3d electrons in the electrode are more tightly bound than the 4s. While transforming a quantity to any other complete basis would give a result with the same number of components N_G (i.e., no information is lost and hence the transforms are invertible), the advantage we sought in a different basis was that key quantities, such as the hamiltonian matrix or the wavefunctions, might be rendered sparse in a systematically exploitable way. Members of our group had previously explored the use of a Wannier basis which rendered the hamiltonian matrix in a sparse form; however, exploiting this sparseness in the calculation of H was problematic, and the benefit of replacing the diagonal T_G and (semi) local V_R with a sparse $(T + V)_w$ in Wannier space was not clear.

A second option which we explored placed emphasis on reducing the cost of the expensive orthogonalization step. we introduced a systematic method of "basis thinning" or "zero suppression" in which, prior to orthogonalization, each wavefunction is scanned for components which are above a set threshold; an index vector of length $N_{thin} < N_G$ is stored which points only to these significant components, at a cost which scales as $M_\psi N_G$. In the orthogonalization loops, this results in the execution of

$$\psi'_2 = \psi_2 - \langle \psi_1 | \psi_2 \rangle \psi_1$$

in N_{thin} operations, rather than N_G. Thus with an overhead cost of order $M_\psi N_G$, the entire orthogonalization loop is reduced to order $M_\psi^2 N_{thin}$. We implemented this algorithm and found that the orthogonalization time is indeed reduced by applying the thinning mechanism, but the speedup at which the errors in quantities of interest become electrochemically intolerable is only about a factor of two. While this is a significant savings, it is not enough to make a full 3d4s–copper slab calculation affordable.

Next, we considered treating the bulk of the electrode slab as 4s–copper, with a "patch" of 3d4s–copper on the surface near where an additional 3d4s–copper ion would undergo the electron transfer reaction. Such a modification would greatly reduce the number of wavefunctions needed in the calculation, though the large cutoff energy would need to be retained in order to describe the few 3d4s–coppers accurately. This higher cutoff energy, however, could be exploited to use a truly first–principles pseudopotential for the 4s–copper (rather than the semi–emprically adjusted pseudopotential used in our group's previous studies). The cost, estimated at a factor of 20 more expensive than the earlier 4s–copper calculations, is on the edge of affordability. While a similar approach has been taken in small–cluster calculations of water adsorbtion on copper (24) , many theoretical issues need to be addressed in order for this to succeed. Included among these are evaluating the work function of the first–principles 4s–copper slab, assessing the impact of the patch on the work function, and estimating the quality of the interaction of the reacting ion with the patch, relative to the true interaction with a fully–described electrode.

As an initial test, we calculated the vacuum work function (defined in simulation as the difference between the electric potential in the vacuum region and the fermi level of the slab) of a 5x5x5 4s–copper slab exposing the (100) face, using ab initio pseudopotentials (22) at the cutoff energy required for 3d4s–copper convergence. The result at 5.8 eV disagrees with experimental results (25) of 4.5 eV by a large amount, indicating that, as feared, the 3d electrons play a significant role at the vacuum interface.

Rather than retrace the path of adjusting the pseudopotential to fit the work function as in reference 10, we considered more radical options to speeding up the calculation. A method which we developed based on the work of Goedecker and Teter (*26*) (with tight–binding electronic structure calculations) yields the ground state energy and electron density by applying the identity

$$\rho(x) = \sum_\lambda f\left(\varepsilon_\lambda\right)\langle x|\lambda\rangle\langle\lambda|x\rangle$$

$$= \sum_\lambda \langle x|f\left(\beta\left(H-\mu\right)\right)|\lambda\rangle\langle\lambda|x\rangle$$

with energy eigenstates $\{|\psi\rangle\}$, position eigenstates $\{|x\rangle\}$, and

$$f(x) = \frac{1}{e^x+1} = \sum_n^{M_H} a_n x^n$$

is the fermi function evaluated as a matrix polynomial. (In practice this is evaluated using Chebyshev polynomials.) In a model with purely local pseudopotentials, the density at all positions x can therefore be evaluated as

$$\rho(x) = \langle x|f\left(\beta\left(H-\mu\right)\right)|\lambda\rangle$$

in $M_H N_G^2 \log N_G$ operations. Equivalently we can write in our plane–wave basis $\{|k\rangle\}$

$$\rho(x) = \sum_k \langle x|f|k\rangle\langle k|x\rangle$$

and introduce a vector of random phases

$$P = \sum_k e^{i\phi_k k}|k\rangle$$

to obtain

$$\langle x|f|P\rangle\langle P|x\rangle =$$

$$\sum_k \langle x|f|k\rangle\langle k|x\rangle + \sum_{k\neq k'} e^{i(\phi_k-\phi_{k'})}\langle x|f|k'\rangle\langle k|x\rangle$$

$$= \rho(x) + \delta\rho(x)$$

Hence summing over a sufficient number M_P of random vectors should give cancellation in $\delta\rho$ due to the randomness of the phases; indeed, self–averaging keeps $\delta\rho$ relatively small in a single vector. The energy can be evaluated as

$$E = V\left(\rho(x)\right) + 1/M_P \sum_P \langle P|T|P\rangle.$$

Advantages of the method include the scaling of the number of operations in the case of local pseudopotentials as $M_H M_P N_G \log N_G$ which in particular is independent of the number of wavefunctions M_ψ and nearly order N_G. The method is easily parallelized (each processor generating a small number of approximate densities), and has low memory requirements. However, trials of the method required values on the order of $M_H \sim 10^3$ and $M_P \sim 10^2$ with $N_G \sim 10^5$, which if parallelized using $M_{cpu} \sim 10^2$ processors gives an estimated time of $10^{3+2+5-2} = 10^8$, compared with the current algorithm at $M_\psi^2 N_G$ (orthogonalization step) which is $10^{2 \times 2+5} = 10^9$; hence, this method also does not yield the required speedup. Evaluation of non–local pseudopotentials will significantly slow the hamiltonian multiplications in the evaluation of the fermi function. In addition, the process must be repeated with a number of chemical potentials to run with fixed electron number, further reducing its utility.

Hence, the daunting task of describing the large number of electronic wavefunctions at the spatial resolution required for accurate structural and energetic calculations poses
a severe challenge to current implementations of density functional theory. While our attempts to utilize the direct dynamics methods in the study of cuprous–cupric electron transfer have not yet met with success, we continue to explore many options and refine current methods.

Aluminum

While we are unable at this time to obtain a more complete description of the copper/water interface, it is a natural next step to consider one of the simpler sp metals. While our collaborator, David Price, has treated the cadmium/water interace, we have chosen for further study aluminum, which has 3 valence electrons per atom (electronic configuration $3s^2\, 3p^1$), due to its widespread use in industrial applications. We are also particularly interested in describing the complete series of interfaces in the aluminum metal / aluminum oxide / water system which is predominant in real–world applications. Though the electronically disorded structure of the oxide is not well suited for description by the direct dynamics code used here, studies of the metal/vacuum and metal/water interface are useful for our development of tight–binding methods for use in treating the oxide–containing interfaces (27).

We considered first a simulation cell containing 125 aluminum centers arranged in a 5x5x5 atom slab of FCC crystal structure exposing the 100 face to vacuum. Cell dimensions of 27.08 x 27.08 x 45.00 au were chosen to match the bulk lattice constant of aluminum metal. The aluminum centers were described with Troullier–Martins pseudopotentials (22) with a cutoff energy of 16 Ry for the plane–wave basis. Relaxation of the electronic density yielded a work function, defined as the difference between the electrostatic potential in the center of the vacuum region and the fermi energy of the slab, of 4.49 eV, which compares well with the experimental value of 4.41 eV (*28*). The binding energy, defined as $E_{slab}/N - E_{atom}$ (and hence including surface energy contributions), is 3.65 eV (3.42 eV experimental (*28*)).

We next proceeded to allow the atomic centers of the aluminum slab to relax. Relaxation attempts with the copper slab, in which a semi–empirical pseudopotential was used, displayed large relaxations corresponding with the unsatisfactory energetics and geometry of small clusters of those pseudoatoms, which we have attributed to the neglect of the 3d copper electrons. We found, in contrast, that the Al_2 dimer using the Troullier–Martins pseudopotentials exhibited a binding energy in quite reasonable agreement with experiment (1.77 eV vs. 1.79 eV experimental (*30*)), but with a separation of 2.63 Å, somewhat larger than the expected 2.47 Å (*30*); hence, we expect the aluminum slab to expand to some extent. Since this was the first attempt at relaxation of the atomic centers using non–empirical pseudopotentials, we used the same supercell dimensions as in the case of the unrelaxed slab. In this case the expansion is constrained to the z direction, and hence its effects are exaggerated (Further studies varying the x– and y–dimensions of the simulation cell to allow symmetric relaxation of the slab are currently underway; we expect the geometric and energetic results to be bounded by those of the unrelaxed slab and the slab relaxed within the fixed supercell).

The slab was relaxed until the average force on the atomic centers was less than 0.01 eV/Å, with the maximum force less than 0.02 eV/Å. The relaxed slab exhibited a binding energy as defined above of 3.71 eV, with the outermost atomic layer displaced 0.18Å in the z direction from the position with bulk lattice structure, consistent with the error in the dimer geometry. The work function was reduced to 4.08 eV.

We have also begun a study of the aluminum slab in water using the water model described in reference 10. We introduced 180 water molecules into the simulation cell in an ice structure and have allowed them to thermally equilbrate to 300K with the electronic structure of the aluminum slab frozen. We have proceeded to allow the electronic structure to follow the motion of the water and look forward to allowing dynamics of the aluminum centers themselves.

Conclusions

We have studied the electrode/electrolyte interface using direct dynamics simulations. Inclusion of the electronic structure of the electrode has revealed qualitative differences in comparison with classical molecular dynamics simulations, in particular with regard to the capacitance of the interface and the character of electric fields near the electrode surface. Our studies of the copper/water interface have shown that, while it is possible to successfully treat the electronic structure, the expense of such calculations currently prohibits the complete study of transition metals and their oxides, due to the large required number of wavefunctions and the high plane–wave cutoff energy necessary to resolve the wavefunctions. We have examined a number of alternative approaches to the electronic structure aspects in order to minimize this expense, but with limited success. However, we have been able to extend our studies to sp metals such as Aluminum, which, with the implementation of first–principles pseudopotentials, are allowing us to explore the effects of electrode nuclear relaxation and, in the near future, dynamics.

Acknowlegements

This work was supported in part by the National Science Foundatio, grant number DMR–952228, by the Department of Energy Office of Science, Materials Science Division grant number DE–FG02–91–ER45455 and by the University of Minnesota Supercomputing Institute.

Literature Cited

1. Walbran, S.; Mazzolo, A.; Halley, J. W.; Price, D. L.,*J. Chem. Phys.* 109, 8076 (1998)
2. Price, D.; Halley, J. W. *J. Electroanal. Chem.* **1983**, *150*, 347.
3. Halley, J.W.; Johnson, B.; Price, D.; Schwalm, M. *Phys. Rev. B* **1985**, *31*, 7695.
4. Halley, J. W.; Price, D. *Phys. Rev. B* **1987**, *35*, 9095.
5. Price, D.; Halley, J. W. *Phys. Rev B* **1988**, *38*, 9357.
6. Schmickler, W.; Henderson, D. *J. Chem. Phys.* **1986**, *85*, 1650.
7. Goodisman, J. *J. Chem. Phys.* **1989**, *90*, 5756.
Amokrane, S.; Russier, V.; Badiali, J.P. *Surf. Sci.* **1989**, *217*, 425.
8. Kornyshev, A.A. *Electrochim. Acta* **1989**, *34*, 1829.
9. Price, D.; Halley, J. W. *J. Chem. Phys.* **1995**, *102*, 6603.
10. Bard, A. J.; Faulkner, L. R. Electrochemical Methods, Fundamentals and Applications, J. Wiley and Sons: NY, 1980; p. 500
11. Halley, J.W.; Walbran, S.; Smith, B. Proceedings of the Workshop on Charge Transfer at ITCP Trieste, ed. Kornyshev, A.; World Scientific, 1997.
12. Stern, O. *Z. Elektrochem.* **1924**, *30*, 508.
13. Toukan, K.; Rahman, A. *Phys. Rev. B* **1985**, *31*, 2643.
14. Hautman, J.; Halley, J. W.; Rhee, Y.–J. *J. Chem. Phys.* **1989**, *91*, 467.
15. Halley, J. W.; Walbran, S.; Price, D. L. Interfacial Electrochemistry, ed. Wieckowski, A. Marcel Dekker: New York, in press.
16. Jackson, J. D. Classical Electrodynamics; J. Wiley and Sons: New York, 1975; Sec. 6.7.

17.   Lecoeur, J.; Bellier, J. P. *Electrochimica Acta* **1985,** *30,* 1027.
18. Haertinger, S.; Doblhofer, K. *J. Electroanal. Chem.* **1995,** *380* , 185.
19. LaGraff, J.R.; Cruickshank, B.J.; Gewirth, A.A. *In .. Nanoscale Physical Properties of Materials..MRS Symp.*; Sarikaya, M.; Wickramasinghe, H.K.; Isaacson, M., Eds. MRS: Pittsburgh, PA, 1994; p.121.
20. Foresti, M. L.; Pezzatini, G.; Innocenti, M.. *J. Electroanal. Chem.* **1997,** *434,* 191.
21. Troullier, N.; Martins, J. L. *Phys. Rev. B* **1991**, *43*, 1993.
22. Walbran, S. *Doctoral Thesis*, University of Minnesota; 1999.
23. Ribarsky, M. W.; Luedtke, W. D.; Landman, U. *Phys. Rev. B* **1985**, *32*, 1430.
24. Lipkowski, J.; Ross, P.N. *Structure of Electrified Interfaces*; VCH Publishers: New York, NY, 1993; p 213.
25. Goedecker, S.;Teter, M. *Phys. Rev. B* **1995**, *51*, 9455.
26. Schelling, P.K. , Halley, J. W., Yu, N. *Phys. Rev. B* **1998,** *51,*1279
27. Lide, D. R., ed., *Handbook of Chemistry and Physics*, 77th ed. Chemical Rubber Cleveland, 1997, p10–214
28. Cox, J. D.; Wagman, D. D.; Medvedev, V. A. *CODATA Key Values for Thermodynamics*; Hemisphere Publishing Corp.: New York, 1989.
29. Huber, K. P.; Herzberg, G. *In NIST Chemistry WebBook, NIST Std. Ref. Database 69*; Mallard, W.G.; Linstrom, P.J., Eds. NIST: Gaithersburg, MD, 2000; (http://webbook.nist.gov)

Chapter 3

Electronic Properties at a Metal–Solution Interface as Viewed by Solid-State NMR

YuYe Tong, Cynthia Rice, Eric Oldfield, and Andrzej Wieckowski

Department of Chemistry, University of Illinois at Urbana-Champaign, 600 South Mathews Avenue, Urbana, IL 61801

We report our recent experimental and theoretical progress in our electrochemical NMR (EC-NMR) investigations of metal/solution interfaces. By correlating EC-NMR with *in-situ* infrared spectroelectrochemistry of chemisorbed CO, we discuss in detail, at an electronic level, the CO-metal (Pt) bonding as well as the electrode potential Stark effect. The ability to carry out the detailed EC-NMR investigations and couple it with other *in-situ* spectroscopic techniques opens a new promising avenue of research in the discipline.

Introduction

Using the recently developed solid-state NMR/electrochemical technique[1-9], it is now possible to investigate metal/liquid interfaces under potential control[3,9], to deduce electronic properties of electrodes (platinum)[6] and of adsorbates (CO)[7], and to study the surface diffusion of adsorbates[1,3,5]. The method can also provide valuable information on the dispersion of commercial carbon-supported fuel cell platinum electrocatalysts and on electrochemically-generated sintering effect[8].

The uniqueness of the EC-NMR approach is that it combines the strengths of two important fields. On the one hand, solid-state NMR possesses a unique technical versatility, elemental specificity, and local chemical/electronic environmental sensitivity. Therefore, information on many important structural, electronic, and dynamical properties at electrode surfaces can be obtained. On the other hand, modern interfacial electrochemistry[10] provides tremendous materials-engineering capabilities, in which surface/interfacial compositions, and thence their physical and chemical properties, can be engineered in a well-controlled fashion, by electro-dissolution or electro-deposition, usually at room temperature. In addition, large, tunable electric fields (up to 10^9 V/m) can be generated at the electrode/electrolyte interface. When compared to non-electrochemical interfaces, the ability to vary the electric field, through varying the electrode potential, offers an additional degree of freedom with which to investigate various aspects of interfacial structure. Solid state NMR therefore represents, as we will demonstrate below, a powerful new approach to obtaining a comprehensive understanding of the basic physical and chemical processes occurring at the electrochemical interface.

In the spirit of the symposium, we will discuss below the essential experimental and theoretical NMR methods involved in investigating metal-CO bonding, and we will demonstrate how the electronic properties of the interface can be portrayed by NMR studies of both substrate (Pt) and adsorbate (CO) species. We will also show that such results can be used to provide an electronic interpretation of *in-situ* infrared results[11], and how they substantiate the Blyholder (5σ forward- and $2\pi^*$ back-donation) bonding mechanism[12] between CO and a transition metal surface.

Theoretical Background: Two-Band Models for ^{195}Pt and ^{13}C NMR of Platinum Substrate and CO Adsorbate

There are two unique features of NMR in a metal: the presence of the Knight shift[13], K, and the Korringa spin-lattice relaxation mechanism[14], $1/T_1 \propto T$, where T is the absolute temperature at which the relaxation rate is measured. Both K and $1/T_1$ are directly related to the Fermi level local density of states (E_f-LDOS) at the observed nucleus through the hyperfine interactions between nuclear and conduction electron spins. Both are described by the same Hamiltonian, which includes Fermi contact, nuclear spin-electron spin dipole-dipole, and spin-orbital contributions[15]:

$$H = \gamma_n \hbar \mathbf{I} \bullet \mu_B \{\frac{8\pi}{3}\mathbf{S(r)}\delta(\mathbf{r}) - [\frac{\mathbf{S}}{r^3} - \frac{3r(\mathbf{S}\bullet\mathbf{r})}{r^5}] - \frac{l}{r^3}\} \tag{1}$$

Here, μ_B is the Bohr magneton, γ_n is the nuclear gyromagnetic ratio, $\mathbf{I}$, $\mathbf{S}$, and l the nuclear spin, electron spin, and electron orbital moments, respectively; and $\mathbf{r}$ is the radius vector of an electron with respect to the nucleus as the origin. When detailed band structure at the Fermi level cannot be neglected, the electrons in different bands will contribute differently to K and $1/T_1$[16] via eq.1. Only the sum of these contributions, however, can be determined experimentally. Since s and d bands for platinum[17] and 5σ and $2\pi^*$ bands for chemisorbed CO[18] cross the Fermi level, phenomenological two-band models, which neglect inter-band interactions, have been devised for obtaining the E_f-LDOS values at platinum surfaces[19], and at the carbon atom of chemisorbed CO[7]. The model can be expressed for platinum by the following equations:

$$K = K_s + K_d + K_{orb}$$
$$= \mu_B D_s(E_f)H_{hf,s}/(1-\alpha_s) + \mu_B D_d(E_f)H_{hf,d}/(1-\alpha_d) + \chi_{orb}H_{orb}/\mu_B \tag{2}$$
$$S(T_1T)^{-1} = k(\alpha_s)K_s^2 + k(\alpha_d)K_d^2 R_d + [\mu_B D_d(E_f)H_{hf,orb}]^2 R_{orb} \tag{3}$$

and for chemisorbed CO:

$$K = K_{5\sigma} + K_{2\pi^*} + K_{orb}$$
$$= \mu_B D_{5\sigma}(E_f)H_{hf,5\sigma}/(1-\alpha_{5\sigma}) + \mu_B D_{2\pi^*}(E_f)H_{hf,2\pi^*}/(1-\alpha_{2\pi^*}) + K_{orb} \tag{4}$$
$$S(T_1T)^{-1} = k(\alpha_{5\sigma})K_{5\sigma}^2 + k(\alpha_{2\pi^*})K_{2\pi^*}^2/2 + (13/5)[\mu_B D_{2\pi^*}(E_f)H_{hf,orb}]^2 \tag{5}$$

Here, S is the Korringa constant $(\gamma_e/\gamma_n)^2(\hbar/4\pi k_B)$, and χ_{orb} is the orbital susceptibility. And $D_{s/5\sigma}(E_f)$ and $D_{d/2\pi*}(E_f)$ are the separate $s/5\sigma$- and $d/2\pi*$-like densities of the states at the Fermi level. $H_{hf,l}$, with $l = s/5\sigma$, $d/2\pi*$, and orb, stand for $s/5\sigma$ -band, $d/2\pi*$-band, and orbital contributions. For platinum, R_d and R_{orb} are the reduction factors due to the orbital degeneracy at the Fermi level[20,21]:

$$R_d = [\frac{f^2}{3} + \frac{(1-f)^2}{2}] \text{ and } R_{orb} = \frac{2f}{3}(2 - \frac{5f}{3}),$$

where $f = 0.7$ is the relative weight of the t_{2g}- and e_g-type d orbitals at the Fermi energy. The electronic many-body effects may be taken into account through the Stoner enhancement factor in the Knight shift and the Shaw-Warren de-enhancement factor in the spin-lattice relaxation rate. The $s/5\sigma$- and $d/2\pi*$-like Stoner factors are therefore introduced as $\alpha_l = I_l D_l(E_f)$ with $l = s/5\sigma$, $d/2\pi*$, where $I_{s/5\sigma}$ and $I_{d/2\pi*}$ are exchange integrals. The Shaw-Warren de-enhancement factor $k(\alpha)$ is assumed to be given by the Shaw-Warren relationship[22], both for $s/5\sigma$-like and $d/2\pi*$-like electrons. In our work, the $k(\alpha)$ is approximated by polynomials up to eighth order:

$$k(\alpha) = 1.0 + \sum_{i=1}^{8} a_i \alpha^i,$$

where a_i , Table I, were determined by numerically fitting the Shaw-Warren curve[22].

Table I. The polynomial coefficients for $k(\alpha)$.

i	1	2	3	4	5	6	7	8
a_i	-0.7491	-0.1518	-0.7251	3.283	-8.449	11.80	-8.496	2.493

All of the hyperfine parameters used in eqs. 2, 3, and 4, 5, which are considered constants, are listed in Table II. Using these parameters and the experimentally determined K and $S(T_1T)^{-1}$, eqs. 2 and 3 (4 and 5) become a pair of coupled equations

Table II. Parameters used for Ef-LDOS analysis.

	K_{orb} (ppm)	$I_{s/5\sigma}$ $(Ry)^b$	$I_{d/2\pi*}$ $(Ry)^b$	$H_{hf,s/5\sigma}$ (kG)	$H_{hf,d/2\pi*}$ (kG)	$H_{hf,orb}$ (kG)
Pt^a	2100	0.098	0.037	2700	-1180	1180
CO^a	160	0.086	0.078	363	17	50

^{a}For detailed entries of the parameters, see Reference 19 for Pt and Reference 7 for CO; ^{b}1 Ry = 1 Rydberg =13.6 eV.

for two unknowns: $D_{s/5\sigma}(E_f)$ and $D_{d/2\pi*}(E_f)$. By solving the equations, one can obtain the numerical values for the E_f-LDOS at the observed nuclei. Although one needs to exercise some caution in that the absolute values so obtained may contain some systematic errors, due to the approximate nature of the models used, the physical picture, i.e., the increase/decrease in $D(E_f)$ simply reflects the variation in K and $S(T_1T)^{-1}$, which is of course *independent* of any theoretical model.

Experimental

The electrode materials used were carbon-black (XC-72)-supported commercial fuel-cell-grade nanoscale platinum electrocatalysts (E-TEK Inc., Natrick, MA). Five samples having average particle sizes (provided by E-TEK) of 2.0, 2.5, 3.2, 3.9, and 8.8 nm, as determined by x-ray diffraction, were investigated. To electrochemically characterize the Pt surfaces, cyclic voltammetry (CV) was performed at a sweep rate of 20 mV/min in a conventional three-electrode flow-cell, containing a platinum gauze counter electrode, a 1 M NaCl Ag/AgCl reference electrode (to which all the potential values in this paper were referred), and a working electrode consisting of supported platinum catalyst particles (using 50 to 380 mg) contained within a platinum boat, under a blanket of ultra pure argon. The cell potential was controlled by a PGP201 Potentiostat/Galvanostat manufactured by Radiometer (Villeurbanne). The electrolyte used was 0.5 M H_2SO_4 (Mallinckrodt Baker Inc.) diluted with Millipore water (Millipore Corp.). NMR samples were prepared by initially cleaning the Pt surface (approximately 500 mg), then holding the cell at 250 mV for 1 to 2 hours, until the current decreased to insignificant levels. Then, surface ^{13}CO was produced on the electrode from the electrochemical dissociative chemisorption of 99% ^{13}C labeled methanol (Cambridge Isotopes Laboratories) in 0.5 M H_2SO_4, at a potential of 0 mV. After the electrochemical treatment, the platinum electrode material, together with a small portion of supporting electrolyte which protected the surfaces, was transferred into a 10-mm diameter x 25 mm length tube and flame-sealed under reduced pressure, in a nitrogen atmosphere. Notice that although without direct potential control, the NMR sample was immersed in the electrolyte, therefore had normal metal/electrolyte interfaces.

All NMR measurements were carried out on "home-built" NMR spectrometers equipped with either 3.5-inch bore 8.47 or 14.1 Tesla (T) superconducting magnets (Oxford Instruments), Aries data acquisition systems (Tecmag), and (at 8.47 T) using an Oxford Instruments CF-1200 cryostat. A Hahn spin-echo pulse sequence ($\pi/2$—τ—π—τ—acquisition) with 16-step phase cycling to eliminate ringdown, was used for data acquisition. The typical $\pi/2$ pulse length was between 5 to 16 µs, depending on experimental conditions. Chemical (or frequency) shifts for ^{13}C are given in ppm from tetramethylsilane (TMS). All spin-lattice relaxation times were measured by using an inversion-recovery method for ^{13}C (at spectral maximum) and a saturation-recovery for ^{195}Pt (at maximum of surface peak), followed by a Hahn-echo acquisition sequence. Notice that both the ^{195}Pt and ^{13}C NMR spectra are very inhomogeneously broadened[6, 7] and, therefore, the spin diffusion is quenched. Thus, what measured in these experiments were the representative *local* spin-lattice relaxation rates at the carrier frequency.

^{195}Pt NMR of Nanoscale Carbon-Supported Platinum Electrodes

As we have demonstrated previously, carbon-supported nanoscale platinum electrodes retain quite closely the ^{195}Pt NMR spectral characteristics of isolated small platinum particles[6], with the clean surface platinum atoms resonating at 1.100 G/kHz with respect to the value for bulk atoms, 1.138 G/kHz. This corresponds to a difference of ca. 3 kG in the local magnetic field seen by the platinum atoms, at 8.47 T. In Figure 1A-C, typical ^{195}Pt NMR electrochemical cleaning, and the layer-model deconvolution of the clean surface spectrum (Figure 1C), are shown. The surfaces of the as-received materials, which were covered by O and/or OH species (indicated by the peak at 1.098 G/kHz), were effectively cleaned by holding the electrode potential within the electrochemical double-layer region (250 mV). This mild surface reduction procedure should be contrasted to the rigorous and often technically demanding methods employed at the solid/gas interface, which usually involves several cycles of high temperature calcination and reduction, limiting the choice of catalyst support to, typically, non-conductive oxides such as alumina, titania, and silica.

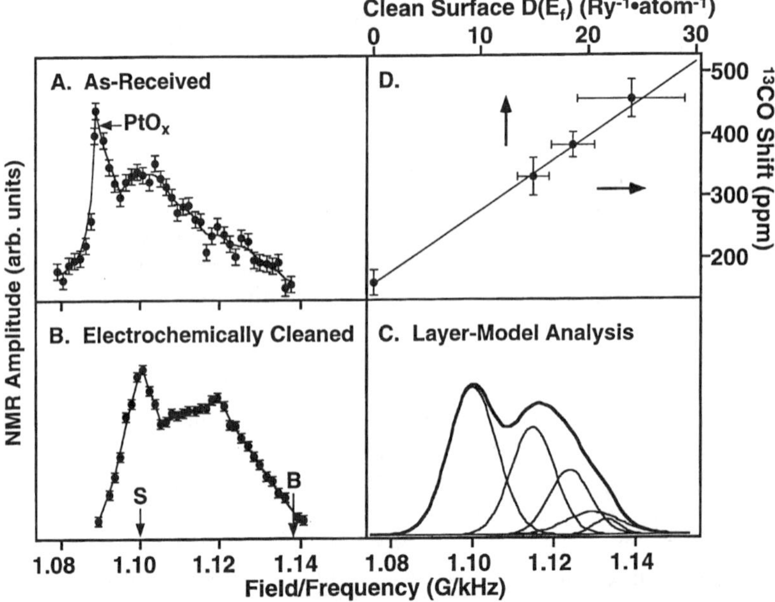

Figure 1. *Typical ^{195}Pt NMR spectra showing electrochemical cleaning and a layer-model analysis of the 2.5 nm sample: **A**, As received catalyst; **B**, electrochemically cleaned in 0.5 M H$_2$SO$_4$ and **C**, layer-model deconvolution of spectrum **B** (see Text for details); **D**, correlation between the Fermi level total electronic densities of states for clean surface Pt (in Ry^{-1}•atom^{-1}) and the ^{13}CO Knight shift.*

The NMR layer model assumes that the nanoscale platinum particles can be represented by ideal cubo-octahedral particles built up layer-by-layer from the central atom, that NMR signals from atoms within a given layer can be approximated by a Gaussian, and that the average Knight shift of the nth layer, K_n, which is the center of the corresponding Gaussian, "heals" back exponentially towards the bulk platinum position when moving inwards. That is: $K_n = K_\infty - (K_0 - K_\infty)\exp(-n/m)$, where, n is the layer number, counting inwards from the surface layer (where $n = 0$), m is the characteristic number (of layers) defining the "healing length" for the Knight shift, and K_0 and K_∞ are the Knight shifts of the surface layer and the bulk, respectively. The relative contribution of each Gaussian is dictated by the fraction of the atoms within the corresponding layer, which can be determined from the size distribution of the sample. The healing length, defined as m times the distance between two consecutive layers (0.229 nm for Pt), is 0.46 nm ($\cong$ 2 Pt layers) from the deconvolution shown in Figure 1C, which is to be compared with 0.31 nm ($\cong$1.35 Pt layers) in the gas phase case[23], indicating a significant change in the electronic structure of the carbon-supported sample.

Since there is a clear NMR discrimination of the surface peak at 1.100 G/kHz, we can obtain the clean surface E_f-LDOS by using detailed relaxation measurements[6], and the value of $S(T_1T)^{-1}$, which is needed to deduce the $D(E_f)$'s by solving eqs. 2 and 3. For the sample shown in Figure 1B, we found that $D_s(E_f) = 5.1$ Ry^{-1}•atom^{-1} and $D_d(E_f) = 13.5$ Ry^{-1}•atom^{-1}, values which are to be compared to the corresponding results in the gas phase case, 3.9 and 10.9 Ry^{-1}•atom^{-1}, respectively[19]. Apparently, there is a significant enhancement in the E_f-LDOS in the electrocatalyst samples. We also found that when CO was adsorbed onto transition metal surfaces, the ^{13}C NMR shift of CO responds linearly to the clean surface E_f-LDOS before adsorption. The data[6] are graphically represented in Figure 1D. The straight line is a linear fit to the data, giving a slope of 12 ppm/Ry^{-1}•atom^{-1}, a key relationship which can be substantiated from a completely different data set, which provides an independent estimate of the slope, confirming the above results[6]. In particular, it has previously been found that for CO on Pt surfaces, there exists a linear relationship between the infrared vibrational stretching frequency[24], as well as the ^{13}C shift[3,25], and the externally applied electrical potential (in an electrochemical cell). It has also been shown that there is a linear relationship between the IR vibrational stretching frequency and the clean surface E_f-LDOS in the gas phase[26]. When combined, these results lead to a linear ^{13}C shift/E_f-LDOS, correlation. The slopes are respectively 30 cm^{-1}/V, -71 ppm/V, and -4 cm^{-1}/Ry^{-1}•atom^{-1}, from which one can readily deduce a slope of 9.5 ppm/Ry^{-1}•atom^{-1}. This is in excellent agreement with the slope of 12 ppm/Ry^{-1}•atom^{-1} found in Figure 1D. The close agreement, (within experimental error) of these two data sets, strongly indicates that the potential dependence of the ^{13}C shift, as well as the IR vibrational stretching frequency of adsorbed CO, are dominated by changes in the E_f-LDOS at the metal/solution interface. This conclusion is further substantiated by ^{13}C NMR and *in-situ* infrared results on chemisorbed CO, as described below.

The linear relationship found in Figure 1D is important for at least two reasons. First, it demonstrates the validity of the frontier orbital-interaction picture of metal surface chemistry, in which the importance of the clean surface E_f-LDOS is highlighted[27]. Second, it puts ^{13}C NMR spectroscopy of chemisorbed CO on a firm footing for probing the electronic properties of transition metal surfaces *before* CO chemisorption.

^{13}C NMR of CO Chemisorbed onto Nanoscale Carbon-Supported Platinum Electrode Surfaces

We show in Figure 2 a variety of ^{13}C NMR and electrochemical results on our Pt-^{13}CO systems. First, Figures 2A-C show ^{13}C NMR spectra of CO chemisorbed onto 2.0, 2.5, 8.8 nm samples; second, Figures 2D-F demonstrate the temperature dependence of the spin-lattice relaxation rate measured at the peak positions, and third, Figures 2G-I present the corresponding cyclic voltammograms (CV) of the platinum catalyst surfaces. The disappearance of the voltamemetrically resolved features within the hydrogen evolution region (<50mV) as particle size decreases indicates that the surface inhomogeneity increases, which is also reflected in the broadening of the ^{13}C NMR linewidth. The peak at ca. 170 ppm arises from the graphitized carbon support. It is clearly discriminated, however, from chemisorbed

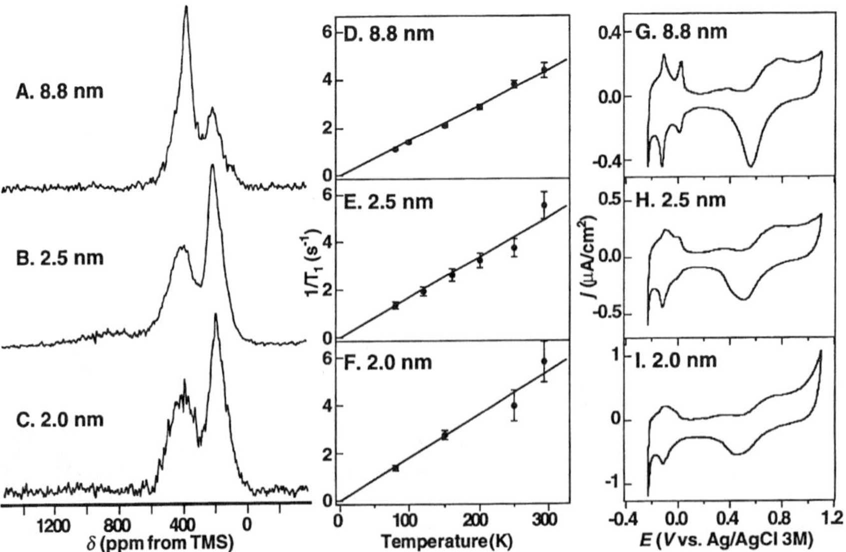

Figure 2. ***I.*** *14.1 T ^{13}C NMR spectra of chemisorbed CO on: **A**, 8.8 ; **B**, 2.5; and **C**, 2.0 nm particle size Pt-carbon (E-TEK, Inc.) samples. **II.** Temperature dependences of the ^{13}C NMR spin-lattice relaxation rate at 8.45 T for chemisorbed CO, measured at peak positions in A-C: **D**, 8.8; **E**, 2.5; and **F**, 2.0 nm particle size samples. The straight lines, which pass through the origin, are the fits to the data, indicating Korringa relationships, and give the $T_1 T$ values shown in Table 4. **III.** Cyclic voltammograms of the corresponding samples: **G**, 8.8; **I**, 2.5; and **H**, 2.0 nm samples.*

CO by virtue of its spin-lattice relaxation behavior. All chemisorbed ^{13}CO's obey the Korringa relationship: T_1T = constant, as evidenced by the straight lines obtained from the $1/T_1$ vs. T plots, Figures 2D-F. These results indicate that spin-lattice relaxation is dominated by conduction electron spin fluctuations at the respective ^{13}C nuclei. The values for T_1T, obtained from a linear fit to the data, are shown in Table III. The data points at 293 K were taken at a magnetic field of 14.1 T, while all other points were obtained at 8.47 T. The fact that the 14.1 T points follow the same $1/T_1$ vs T line also points to the presence of conduction electron spin fluctuations at the ^{13}C sites, since Korringa relaxation is field independent. We also list in Table III the data for CO coverage, ^{13}C peak position and the linewidth obtain by using a Gaussian/Lorentzian fit to the peak of the spectrum. As can be seen from Table III and Figure 2, the ^{13}CO NMR results are particle size dependent. In particular, the peak broadens, and the T_1T decreases, as the particle size decreases. The broadening of the spectra reflects the diversity of surface sites available to Pt-CO bonding, which is indicated by both the change in the line shape of cyclic voltammograms within the hydrogen evolution region, and the T_1 anisotropy across the spectrum[7]. On the other hand, the overall decrease in T_1T indicates the change in the electronic properties of the substrate because, as we have shown above, the NMR characteristics of ^{13}CO are directly related to the surface E_f-LDOS of the substrate.

Table III. ^{13}C NMR data and $5\sigma/2\pi^*$ E_f-LDOS for C) ex. MeOH on carbon-supported nanoscale Pt electrodes

Size (nm)	Coverage (%)	Peak, K (ppm)	FWHM (ppm)	T_1T (s•K)	$D_{5\sigma}(E_f)$ (Ry^{-1}•ml^{-1})	$D_{2\pi^*}(E_f)$ (Ry^{-1}•ml^{-1})
8.8[a]	65	351	102	68 ± 2	0.6 ± 0.1	6.5 ± 0.1
2.5[a]	56	386	174	59 ± 3	0.7 ± 0.2	6.7 ± 0.2
2.0[a]	64	376	245	55 ± 4	0.6 ± 0.2	7.4 ± 0.3
2.5[b]	35	390	177	51 ± 5	0.6 ± 0.2	7.6 ± 0.4

Dissociative chemisorption of MeOH carried out at, [a]0 mV; [b]-150 mV, w.r.t a Ag/AgCl (1M NaCl) reference electrode.

We show in Figure 3 two ^{13}C NMR spectra of CO adsorbed onto the same electrocatalyst, the 2.5 nm sample, but in which the catalytic decomposition of methanol was carried out at different potentials: -150 mV in Figure 3A and 0 mV in Figure 3B (which is the same spectrum as that shown in Figure 2B). The CO coverage, θ, is also different: 0.35 for A and 0.56 for B. The room temperature (293 K) spin-lattice relaxation time for the $\theta = 0.35$ sample, measured at ~390 ppm, was 175 ± 17 ms, giving a $T_1T = 51 \pm 5$ s•K, which is smaller than that of the high coverage sample, $T_1T = 59 \pm 3$ s•K (Table 3), indicating an electronic response of the system to the change in electrode potential.

By inserting the K and T_1T values listed in Table 3 into eqs. 4 an 5, and by solving these equations, we obtain the LDOS partitions between $D_{5\sigma}(E_f)$ and $D_{2\pi^*}(E_f)$, as shown in Table 3. Two important observations can be made from these results. First, in all cases, the contribution from $2\pi^*$-like electrons to the ^{13}C NMR observables is dominant, with $D_{2\pi*}(E_f)$ being about 10 times larger than $D_{5\sigma}(E_f)$. Second, while $D_{5\sigma}(E_f)$ is almost constant in all cases, $D_{2\pi*}(E_f)$ varies substantially. Although NMR cannot provide complete information about the total electron densities, which are the integrals of the LDOS from the bottom of the conduction band to the Fermi level, it precisely measures the electron density at the Fermi level. Since eqs. 4 and 5 can be considered an NMR formulation of the Blyholder 5σ forward- and $2\pi*$ back-donation mechanism for metal-CO bonding, the variation in E_f-LDOS followed by NMR as the physical/chemical environment around the observed nucleus changes, offers useful new quantitative insights into this bonding question.

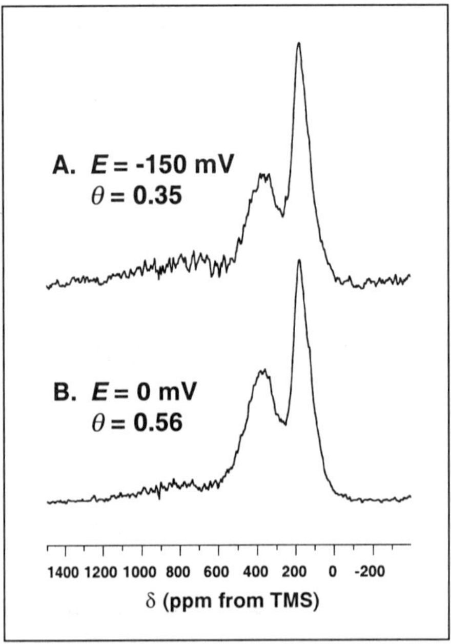

Figure 3. *14.1 T ^{13}C NMR spectra of CO adsorbed onto the same electrocatalyst, the 2.5 nm sample, but the catalytic dissociative chemisorption of methanol was carried out at different potentials: A. −150 mV; B. 0 mV (which is the same spectrum as that shown in Figure 2B). The CO coverage is also different: 0.35 for A and 0.56 for B.*

Now, as is well-known, the 5σ energy level of an isolated CO is far below platinum's d-band, while the $2\pi*$ level is above Fermi level, but with a smaller energy difference. It is therefore expected, following the frontier-interaction picture[27], that after chemisorption, $D_{5\sigma}(E_f)$ should be relatively constant and much smaller than $D_{2\pi*}(E_f)$, as we observe. In addition, there is noticeable variation in $D_{2\pi*}(E_f)$, which strongly suggests that the Fermi level cuts the rising tail of the $2\pi*$ band, as in the case of the CO-Ni system studied by extended Hückel theory[27]. This means that there may exist a monotonic relationship between the degree of back-donation, i.e., the integral from the bottom of $2\pi*$ band to the Fermi level, and the $D_{2\pi*}(E_f)$. That is, the larger the $D_{2\pi*}(E_f)$, the stronger the back-donation, and the weaker the C—O bonding.

Since the weakening of the C—O bond can also be monitored by infrared (IR) spectroscopy, one expects that the IR stretching frequency of adsorbed CO will be smallest for 2.0 nm sample and largest for 8.8 nm sample, as we discuss in the following section.

Correlating ^{13}C NMR with in-situ Infrared Spectroscopy of Chemisorbed CO

It has been previously established that the weakening of the C—O bond upon CO adsorption onto platinum surfaces can be correlated to the surface E_f-LDOS, before adsorption[26]. However, these results did not provide detailed information as to what extent CO's frontier 5σ and $2\pi^*$ orbitals were involved in metal-CO bonding. Since the two-band model, eqs. 4 and 5, allows the partitioning between 5σ and $2\pi^*$ orbitals at the Fermi level to be investigated, we are now in a position to understand the IR results on a more quantitative, electronic basis. In addition, this electronic structural information on metal-CO bonding may offer new insights into the origin of the Stark tuning effect, i.e., the response of the CO stretching frequency to the change in electrode potential (the more positive the electrode potential, the higher the stretching frequency).

In order to test these ideas, we have investigated, using *in-situ* *S*ubtractively *N*ormalized *I*nterfacial *F*ourier *T*ransform *I*nfrared *R*eflectance *S*pectroscopy (SNIFTIRS), five fuel-cell-grade, carbon-supported nanoscale platinum electrode materials, having sizes ranging between 2.0 and 8.8 nm[11]. Three of these, 2.0, 2.5, 8.8 nm, were the same electrocatalysts as those investigated by ^{13}C NMR. The CO stretching frequency at a constant electrode potential, (220 mV), and the Stark tuning rate, as a function of size, are listed in Table IV, and are shown graphically in Figure 4A. At a constant electrode potential, the CO stretching frequency increases with increasing particle size, from 2042 cm^{-1} for the 2.0 nm sample to 2056 cm^{-1} for the 8.8 nm sample. It is well-known that the stretching frequency of chemisorbed CO depends on

Table IV. The particle size dependence of the CO IR frequency and the Stark tuning rate.

Size (nm)	IR frequency (cm^{-1})	Stark tuning rate (cm^{-1}/V)
2.0	2042 ± 2	62 ± 2
2.5	2053 ± 2	53 ± 1
3.2	2059 ± 1	43 ± 1
3.9	2056 ± 1	46 ± 1
8.8	2056 ± 1	46 ± 1

its coverage on metal surfaces. This can vary either due to different initial coverage or due to electro-oxidation of CO, as the electrode potential is moved to more positive values. However, this effect does not appear to influence our results since: (1) the initial CO coverage, determined for all samples by CO stripping current, *via* cyclic voltammetry, showed an average value of 60.9 ± 4.8% of a monolayer; (2) CO stripping take place at a potential around 420 mV, far above 220mV we have used. We therefore conclude that the observed variation in the stretching frequency is not due to a change in CO coverage. Neither is it due to any differences in electrochemical environment, since the same electrolyte was used in each case. Rather our results appear to be originated in differences in the electronic properties of the electrocatalyst, as further demonstrated by the ^{13}C NMR investigations[6,11].

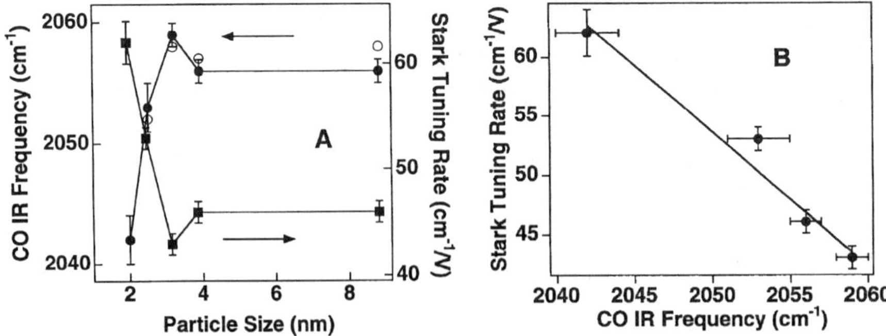

Figure 4. *A, Plot of CO IR stretching frequency measured at 220 mV (left side, filled and open circles with respectively 4 cm^{-1} and 1 cm^{-1} instrumental resolution) and the Stark tuning rate (right side, filled squares) vs. particle size.* ***B***, *correlation between the CO IR stretching frequencies and the corresponding Stark tuning rates. Note that in* ***B*** *the point for the 3.9 nm sample overlaps that for the 8.8 nm sample.*

In contrast to expectations[28,29] from classical electrostatic pictures, our results indicate that the Stark tuning rate, i.e., the potential dependence of the vibrational frequency, is strongly particle-size dependent (Table IV and Figure 4A). The Stark tuning rate is greatest for the 2.0 nm sample, at 62 cm^{-1}/V, while the value of the 8.8 nm sample is 46 cm^{-1}/V, results which should be compared with the value for polycrystalline Pt electrodes of 30 cm^{-1}/V[24]. Interestingly, both the particle size dependence of the CO stretching frequency, and of the Stark tuning rate, show a local irregularity for the 3.2 nm sample, Figure 4A. The reason for this behavior is not understood at this time. However, when the Stark tuning rate is plotted against the IR frequency, using the particle size as an implicit parameter, an almost linear correlation is seen, Figure 4B. Such a linear correlation strongly suggests that the particle size dependence of the CO stretching frequency and of Stark tuning rate have the same origin.

It is also of interest to see how the IR results can be correlated with those obtained in the ^{13}C NMR investigations. The IR stretching frequencies at 0 mV, a potential at which the ^{13}CO's were adsorbed onto the NMR samples, were 2037, 2046, 2051 cm^{-1} for the 2.0, 2.5, 8.8 nm samples, respectively. A linear correlation can be found between these IR values and the corresponding values of $D_{2\pi*}(E_f)$ listed in Table III, with a slope of −15 cm^{-1}/(Ry•molecule)$^{-1}$ and an intercept of 2147 cm^{-1}

(the stretching frequency of free CO is 2143 cm^{-1} [30]). Such a correlation suggests that: (1) weakening of the C—O bond is mainly through the $2\pi^*$ back-donation mechanism and (2) there indeed exists a monotonic relationship between the degree of back-donation, measured by IR, and the $D_{2\pi^*}(E_f)$, measured by ^{13}C NMR. Thus, changes in chemical properties of adsorbed CO are expected to be primarily determined by variations in $2\pi^*$ back-bonding. This reasoning is also applicable to the Stark tuning effect. More specifically, as for the 2.5 nm sample (Table III), the $D_{2\pi^*}(E_f)$ was found to be 6.7 Ry^{-1}•molecule^{-1} when the dissociative chemisorption of methanol was carried out at 0 mV but becomes 7.6 Ry^{-1}•molecule^{-1} when it was done at −150 mV. Thus, the Stark tuning effect can be understood as follows: when the electrode potential is moving in a positive/negative direction, it suppresses/enhances the $2\pi^*$ back-donation, hence increases/decreases the CO stretching frequency.

Conclusions and Perspectives

The relationship between the ^{13}C shift and D_{total} $(E_f) = D_{5\sigma}(E_f)+D_{2\pi^*}(E_f)$ (see Table III), can be described by a linear relationship having a slope of ca. 28 ppm/Ry•molecule, with a R value of 0.997. Since we also know that the slopes of the linear relationship between the ^{13}C shift and the clean surface $D(E_f)$ of platinum, as well as that between the CO IR stretching frequency and the $D(E_f)$ at ^{13}C of CO, are 12 ppm/Ry•atom and -15 cm^{-1}/Ry•molecule, respectively, we can therefore readily deduce a slope of –6 cm^{-1}/Ry•atom for the relationship between the CO IR stretching frequency and the clean surface $D(E_f)$ of platinum. This is in reasonably good agreement with the value of –4 cm^{-1}/Ry•atom previously determined directly from experiment [26]. This agreement clearly supports the idea that the ^{195}Pt NMR, ^{13}C NMR and IR results are all influenced by the same property, the Fermi level local density of states.

Overall then, our results demonstrate that ^{195}Pt NMR of platinum electrocatalysts and ^{13}C NMR of chemisorbed CO offer a powerful pair of surface microscopic probes, from which some detailed electronic properties of the metal/liquid interface can be derived. The internal consistency among all of data (^{195}Pt NMR, ^{13}C NMR, CO IR) provides strong support for the frontier-orbital interpretation of surface chemistry, and of the Blyholder metal-CO bonding mechanism. Using this approach, we will show elsewhere how ruthenium additives to platinum enhance the CO tolerance of these electrodes, which is one of the major challenges for fuel cell science. The scope of NMR electrochemistry can also be expected to broaden as modern NMR instrumentation is pushed toward higher magnetic fields. Higher magnetic fields can be expected to make NMR research possible with low, quadrupolar nuclei such as, for instance, ^{99}Ru (important for fuel cell science) and potentially also for ^{197}Au, of importance in the field of self-assembled-monolayer research, as well as facilitating the observation of other important metals, such as ^{103}Rh and $^{107/109}$Ag.

Finally, it is also instructive to briefly compare and contrast electrochemical metal surface NMR to other in-situ electronic-structure probing techniques, such as x-ray near-edge absorption spectroscopy (XANES) and scanning tunneling spectroscopy (STS). Like NMR, XANES, or x-ray absorption spectroscopy (XAS) in general, is element specific and basically a bulk spectroscopy because of the penetration power of x-ray. Thus like NMR, XANES has to use high-surface-area materials to be surface relavant[31]. XANES is less surface-specific than (^{195}Pt) NMR, because it cannot discriminate the spectral contributions of surface from those of non-surface atoms. More specifically, XANES probes the overall vacant electronic orbitals over about 50 to 100 eV (~ 10^7 K) *above* the Fermi level by exciting electrons to those levels, while metal NMR can measure the exact ground-state electron density *at* the Fermi level. Therefore, the information obtained by these two techniques is highly complementary. STS can in principle probe the local surface electron density at the Fermi level, as metal NMR does, because the tunneling current is proportional to the local electron density at the Fermi level[32]. However, such a spectroscopic approach receives much less attention than it deserves[33]. Since surface-related problems are very complex, a combined approach, whenever possible, is always recommended.

Acknowledgements

This work was supported by the United States National Science Foundation (grant CTS 97-26419), by an equipment grant from the United States Defense Advanced Research Projects Agency (grant DAAH 04-95-1-0581), and by the U.S. Department of Energy under Award No. DEFG02-99ER14993.

References

1. Day, J.B.; Vuissoz, P.-A.; Oldfield, E.; Wieckowski, A.; Ansermet, J.-P. *J. Am. Chem. Soc.* **1996**, *118*, 13046-50.
2. Yahnke, M.S.; Rush, B.M.; Reimer, J.A.; Cairns, E.J. *J. Am. Chem. Soc.* **1996**, *118*, 12250-51.
3. Wu, J. J.; Day, J. B.; Franaszczuk, K.; Montez, B.; Oldfield, E.; Wieckowski, A.; Vuissoz, P.-A.; Ansermet, J.-P. *J. Chem. Soc., Faraday Trans.,* **1997**, *93*, 1017-26.
4. Tong, Y. Y.; Belrose, C.; Wieckowski, A.; Oldfield, E. *J. Am. Chem. Soc.* **1997**, *119*, 11709-10.
5. Tong, Y.Y.; oldfield, E.; Wieckowski, A. *Anal. Chem.* **1998**, *70*, 518A-527A.
6. Tong, Y. Y.; Rice, C.; Godbout, N.; Wieckowski, A.; Oldfield, E. *J. Am. Chem. Soc.* **1999**, *121*, 2996-3003.
7. Tong, Y.Y.; Rice, C.; Wieckowski, A.; Oldfield, E. *J. Am. Chem. Soc.* **2000**, *122*, 1123-29.
8. Rice, C.; Tong, Y.Y.; Oldfield, E.; Wieckowski, A. *Electrochimie Acta* **1998**, *43*, 2825-30.
9. Vuissoz, P.-A.; Ansermet, J.-P.; Wieckowski, A. *Phy. Rev. Lett.* **1999**, *83*, 2457-60.

10. *Interfacial Electrochemistry/Theory, Experiment, and Applications;* Wieckowski, A., Ed.; Marcel Dekker, Inc.: New York, Basel, **1999**.

11. Rice, C.; Tong, Y. Y.; Oldfield, E.; Wieckowski, A.; Hahn, F.; F. Gloaguen, Leger, J.-M.; Lamey, C. *J. Phys. Chem. B* **Submitted.**

12. Blyholder, G. *J. Phys. Chem* **1964**, *68*, 2772-78.

13. Townes, C.H.; Herring, C.; Knight, W.D. *Phys. Rev.* **1950**, *77*, 852-53.

14. Korringa, J. *Physica* **1950**, *XVI,* 601-610.

15. Winter, J. *Magnetic Resonance in Metals*; Clarendon Press: Oxford, 1971.

16. van der Klink, J. J. *Adv. in Catal.* **2000**, *44,* 1-117, Academic Press.

17. MacDonald, A. H.; Daams, J.M.; Vosko, S. H.; Koelling, D. D. *Phys. Rev. B* **1981**, *23*, 6377-98.

18. Hammer, B.; Morikawa, Y.; Norskov, J.K. *Phys. Rev. Lett.* **1996**, *76*, 2141-44.

19. Bucher, J.P. van der Klink, J.J. *Phys. Rev. B* **1988**, *38*, 11038-47.

20. Yafet, Y.; Jaccarino, V. *Phys. Rev.* **1964** *133*, A1630-A1637.

21. Clogston, A. M.; Jaccarino, V.; Yafet, Y. *Phys. Rev.* **1964**, *134*, A650-A661.

22. Shaw, R.W., Jr.; Warren, W.W., Jr. *Phys. Rev B* **1971**, *3*, 1562-68.

23. Bucher, J.P.; Buttet, J.; van der Klink, J.J.; Graetzel, M. *Surf. Sci.* **1989**, *214*, 347-57.

24. Kunimatsu, K,; Seki, H.; Goden, W.G.; II, J.G.G.; Philpott, M.R. *Surf. Sci.* **1985**, *158*, 596-608.

25. Vuissoz, P.-A.; Ansermet, J.-P.; Wieckowski, A. *Electrochimie Acta* **1998**, *44*, 1397-1401.

26. Tong, Y.Y.; Meriaudeau, P.; Renouprez, A.J.; Klink, J.J. v. d. *J. Phys. Chem. B* **1997**, *101*, 10155-58.

27. Hoffman, R. *Rev. Mod. Phys.* **1988,** *60*, 601-628.

28. Lambert, D.K. *Phys. Rev. Lett.* **1983**, *50*, 2106-09.

29. Lambert, D.K. *Solid State Communications* **1984**, *51*, 297-300.

30. Ewing, G.E. *J. Chem. Phys.* **1962**, *37*, 2250-56.

31. McBreen, J.; Mukerjee, S., in *Interfacial Electrochemistry/Theory, Experiment, and Applications*; Wieckowski, A., Ed.; Marcel Dekker, Inc., New York •Basel, 1999, pp 895-914.

32. Feenstra, R. M.; Stroscio, J. A.; Tersoff, J.; Ferri, A. D. *Phys. Rev. Lett.,* **1987**, *58*, 1668-71.

33. Valden, M.; Lai, X.; Goodman, D. W. *Science,* **1998**, *281*, 1647-1650.

Modeling Reaction Rates

Chapter 4

Calculating Adiabatic Potential Energy Surfaces for Electrochemical Reactions

W. Schmickler

Abteilung Elektrochemie, University of Ulm, D–86069 Ulm, Germany

A model Hamiltonian for electron exchange between a metal electrode and an electroactive species in a solution serves as the basis for calculating adiabatic potential energy surfaces for electrochemical reactions. The construction of surfaces for specific systems requires input from molecular dynamics and from quantum-chemical calculations. The adsorption of an iodide ion on a Ag(111) electrode is treated as an example.

The calculation of the potential energy surface for the hydrogen exchange reaction was justly considered a major breakthrough in the study of gas-phase kinetics, and the corresponding contour plot can be found in any major textbook on physical chemistry. Such surfaces greatly help in the understanding and the visualization of reactions, and in a few simple cases, where the calculations are sufficiently accurate, they can even be used to make quantitative predictions for the reaction rate. Obviously, it is highly desirable to calculate similar surfaces for electrochemical reactions, but at a first glance the large number of degrees of freedom involved makes this seem like an impossible task. The problem is, however, greatly simplified by the fact that the classical solvent modes that are coupled to the electron exchange can be lumped into one effective solvent coordinate. If the distance of the reactant from the electrode surface is introduced as a second coordinate, one obtains two-dimensional surfaces, which in simple cases give a good impression of how the reaction proceeds. Additional coordinates can be introduced if required: the position parallel to the electrode surface in order to account for corrugation, quantum coordinates if such modes participate, or bond-breaking coordinates for inner-sphere reactions. In this way one may construct multi-dimensional surfaces that represent all the important aspects of the reaction.

There are several ways to calculate such potential energy surfaces. In this work we review a formalism that we have suggested in the past [1, 2], and present explicit calculations for the adsorption of iodide on a Ag(111) surface at the potential of zero charge. These results are based on more exact calculations than previous attempts to construct the surface for the adsorption of iodide on platinum [1]; in addition, we consider the effective charge of the system on the potential-energy surface, which has not been done before.

Model Hamiltonian and basic equations

Much of the formalism on which our calculations are based has been presented before [1, 2]; we summarize the important equations in order to make this paper self-contained. We consider electron exchange between a metal, whose electronic states are labeled by their quasi-momentum k and spin index σ, and an orbital, labeled a, of a reactant in an electrolyte solution. The exchange involves the reorganization of inner and outer sphere modes, which are represented by a phonon bath. We write our model Hamiltonian as the sum of an electronic and a phononic part; the former is identical to the well-known Anderson-Newns Hamiltonian:

$$H_e = \sum_\sigma \epsilon_a n_{a\sigma} + \sum_{k\sigma} \epsilon_k n_{k\sigma} + U n_{a\sigma} n_{a-\sigma} + \sum_{k\sigma} \left[V_k c_{k\sigma}^+ c_{a\sigma} + V_k^* c_{a\sigma}^+ c_{k\sigma} \right] \quad (1)$$

Here, n and ϵ denote the occupation number and the energy of the indicated state; c^+ and c are the creation and annihilation operators, respectively. U denotes the repulsion between two states with opposite spin on the orbital. The last term in square brackets accounts for electron exchange between the metal and the reactant, and V_k is the corresponding matrix element.

We label the phonon states by ν, and denote by q_ν and p_ν their dimensionless coordinates and momenta. The phononic part of our model Hamiltonian is then:

$$H_{\mathrm{ph}} = \frac{1}{2} \sum_\nu \hbar\omega_\nu \left(p_\nu^2 + q_\nu^2 \right) + (z - n_{a\sigma} - n_{a-\sigma}) \left[\sum_\nu \hbar\omega_\nu g_\nu q_\nu \right] \quad (2)$$

where the last term describes the coupling between the electron transfer and the phonon bath; the g_ν are the coupling constants, and z is the charge number of the reactant when the orbital is empty.

In this work we consider the reorganization of classical solvent modes only; for the adsorption of the iodide ion, which we will treat below, this is a good approximation. It is then possible to replace the manifold of solvent coordinates by a single effective solvent coordinate q^*, which can be normalized such that its values have a physical significance: A value of q^* corresponds to a solvent configuration which would be in equilibrium with a reactant of charge q^* [1]. The coupling parameters V_k and g_ν depend on the distance x of the particle from the electrode surface. The surface that we will present below shows the dependence of the potential energy both on the solvent coordinate q^* and the distance x.

Since motion along the two classical coordinates q^* and x is much slower than electron exchange, we can solve the electronic Hamiltonian for given values of these parameters, making use of the Anderson-Newns theory [3, 4]. The electronic interaction induces a broadening Δ of the adsorbate orbital:

$$\Delta(\omega) = 2\pi \sum_k |V_k|^2 \delta(\omega - \epsilon_k) \tag{3}$$

and a self energy:

$$\Lambda(\omega) = \frac{1}{\pi} \mathrm{Pr} \int \frac{\delta(\omega')}{\omega - \omega'} d\omega' \tag{4}$$

where Pr denotes the principal part. In the following we shall use the *wide-band approximation*, in which the broadening Δ is taken as independent of the electronic energy ω; this is a good approximation when the reactant interacts with a wide electronic band. Since the d band of silver lies well below the Fermi level this approximation should hold in the model calculations presented below.

The two spin states are decoupled in the Hartree-Fock approximation in which the Coulomb term is replaced in the following way:

$$U n_{a\sigma} n_{a-\sigma} \approx U n_{a\sigma} \langle n_{a-\sigma} \rangle + U n_{a-\sigma} \langle n_{a\sigma} \rangle \tag{5}$$

where $\langle \ldots \rangle$ denotes the expectation value. At this stage it is convenient to collect the terms containing the occupation number $n_{a\sigma}$ and introduce the electronic energy:

$$\tilde{\epsilon}_{a\sigma}(q^*) = \epsilon_a + \tilde{\Lambda} + U \langle n_{a-\sigma} \rangle - 2\lambda q^* \tag{6}$$

Here $\tilde{\Lambda}$ contains the chemical shift of the level including image forces, and the interaction λ_f with the fast electronic modes of the solvent; $\lambda = 1/2 \sum_\nu \hbar\omega_\nu g_\nu^2$ is the energy of reorganization for electron exchange. The density of states takes on the form:

$$\rho_\sigma(\omega) = \frac{2\Delta}{\left[\omega - \tilde{\epsilon}_{a\sigma}(q^*)\right]^2 + \Delta^2} \tag{7}$$

and the occupation probabilities $y_\sigma \equiv \langle n_{a\sigma} \rangle$ are given by:

$$y_\sigma = \frac{1}{\pi} \mathrm{arccot} \frac{\tilde{\epsilon}_{a\sigma}(q^*)}{\Delta} \tag{8}$$

where the Fermi-Dirac distribution has been replaced by a step function, and the Fermi level of the metal was taken as the energy zero. Finally, the energy of the system, for a given pair $y_\sigma, y_{-\sigma}$ of solutions, is:

$$E[y_\sigma, y_{-\sigma}] = \sum_\sigma \left\{ \tilde{\epsilon}_{a\sigma} y_\sigma + \frac{\Delta}{2\pi} \ln \frac{\tilde{\epsilon}_{a\sigma}^2 + \Delta^2}{(\tilde{\epsilon}_{a\sigma} - U_c)^2 + \Delta^2} \right\}$$
$$- U y_\sigma y_{-\sigma} + \lambda \left(q^{*2} + 2zq^* \right) \tag{9}$$

46

Here U_c is the bottom of the conduction band, which serves as a lower cut-off energy. The energy depends only weakly on U_c; in the calculations for silver reported below we have taken $U_c = -11.5$ eV, since the bottom of the conduction band lies roughly at 7 eV below the Fermi level [5], and the work function of Ag(111) is 4.46 eV. With the aid of equations (6) to (9) the potential-energy surface can be calculated if the relevant parameters are known as a function of the distance x. Within this formalism, both the adiabatic and the non-adiabatic sections of the surface can be obtained. In this note we will only show the former, which is the branch with the lower energy. However, electron exchange can only occur close to the electrode surface; when the particle is far from the surface it can only move along paths that do not entail an electron exchange [6].

Model calculations for the adsorption of iodide on Ag(111)

As a model system we have chosen the adsorption of an iodide ion on a Ag(111) surface at the potential of zero charge. Koper et al. [7] have calculated the interaction of iodine on several metal surfaces as a function of the distance. For the adsorption on Ag(111) the calculated width can be fitted to an exponential decay of the form:

$$\Delta(x) = \Delta_0 \exp(-\alpha x) \tag{10}$$

with $\Delta_0 = 2.6$ eV and an inverse decay length of $\alpha = 0.86\,\text{Å}^{-1}$; x is the distance from the equilibrium adsorption site, for which the adsorption energy is -2.55 eV. In order to reproduce this adsorption energy within our model we have assumed a chemical shift of $\Lambda = -1.3$ eV and assumed that it decays with the same exponent as the width Δ. Figure 1 shows the resulting energy and the total occupancy $y_\sigma + y_{-\sigma}$ as a function of the distance. The energy increases continuously with the separation; the limiting value at large distances is the difference between the ionization energy of the atom (10.45 eV) and the work function of Ag(111), which is 4.46 eV. The behavior of the total occupancy is more interesting: at short distances the occupancy increases with the separation. In this range the solutions are non-magnetic, i.e. both spin states have the same occupancy. The resulting density of states has its center below the Fermi level; as the separation increases the width becomes smaller, and hence the occupancy larger. At a separation of about 1 Å this behavior changes abruptly: the occupancy drops, and the solution is now magnetic, i.e. the two spins states have unequal occupancy. At long distances only one spin state is occupied, since the iodine atom has one unpaired electron.

In order to calculate the potential-energy surface for adsorption from solution we require the interaction of the ion with the solvent. This was taken from the work of Pecina et al [8]; strictly speaking, these calculations were performed for adsorption on Pt(111), but the free energy ΔG_{sol} of interaction of the ion with the solvent, which is

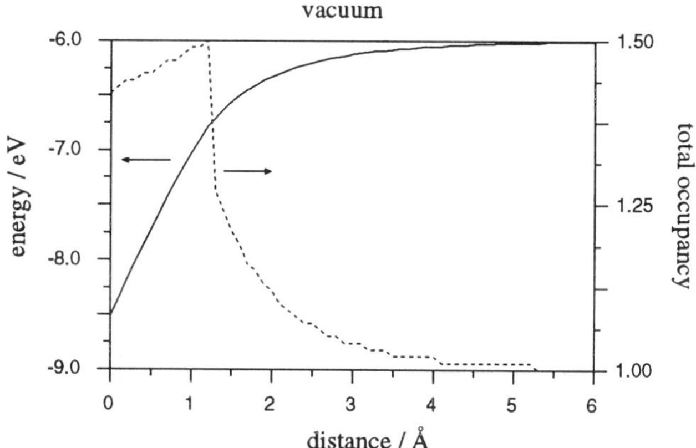

Figure 1. Energy and total occupancy for an iodine atom in vacuum near a Ag(111) surface.

the quantity that we need, depends only weakly on the chemical nature of the metal. So we have taken ΔG_{sol} as a function of the distance, and divided it into two parts: The slow part is caused by the interaction with the solvent dipoles, and gives the reorganization energy $\lambda = (1/\epsilon_\infty - 1/\epsilon_s)\Delta G_{\text{sol}}$, where ϵ_∞ is the optical and ϵ_s the static dielectric constant of water. The fast part is due to the electronic polarization of water, and is given by $\lambda_f = (1 - 1/\epsilon_\infty)\Delta G_{\text{sol}}$; it contributes to the term $\tilde{\Lambda}$ in the electronic energy.

The resulting potential-energy surface is shown in Fig. 2. Far from the electrolyte surface we observe a valley centered at a solvent coordinate of $q^* = -1$ corresponding to the solvated ion; it can be verified from Fig. 3 that the charge $Q = z - y_\sigma - y_{-\sigma}$ on the particle is indeed $Q = -1$ in this region. Near $q^* = -0.3$ there is a ridge in the surface, where the charge changes abruptly to small values. However, there is no second valley corresponding to an unstable atom. In this respect this surface differs from that for the adsorption of iodide on platinum that was shown in [2]. This difference is mainly due to the fact that silver has a lower work function than platinum, although it should also be noted that the present surface is based on more exact calculations. From the shape of the surface it is evident that an electron exchange between the ion and the electrode to form an uncharged iodine atom, which could e occur close to the electrode surface, is highly unfavorable. Such a reaction would occur in the direction of the q^* coordinate, and require a high energy of activation, and would not lead to a stable or even metastable state. However, the ion needs to overcome only a small energy barrier of a few tenths of an electron volts in order to be adsorbed on the electrode surface. Right at the surface, there is a deep valley in which the adsorbate carries a negative partial charge.

The variation of the energy and of the occupancy along the reaction path of minimal energy are shown in Fig. 4. The maximum near 2.7 Å is caused by the partial stripping

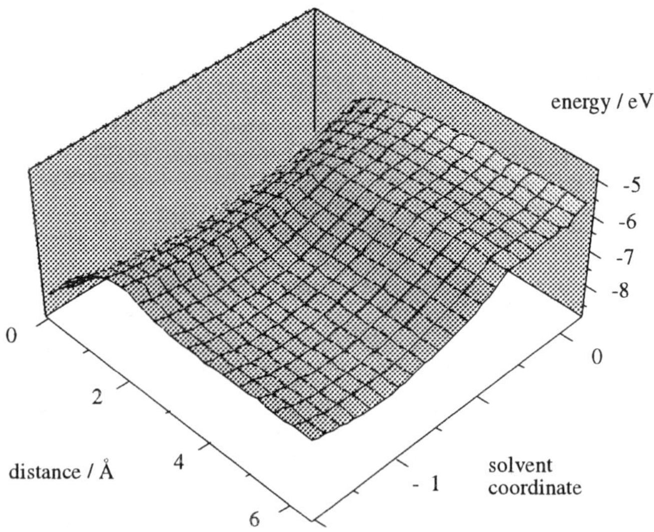

Figure 2. Potential-energy surface for the iodide/iodine system near a Ag(111) electrode.

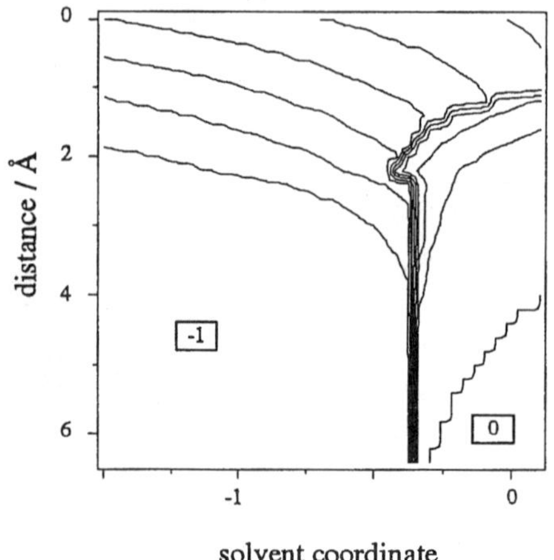

Figure 3. Contour plot of the total occupancy of the valence orbital for the energy surface shown in Fig. 2. Contour lines have been drawn in steps of 0.1.

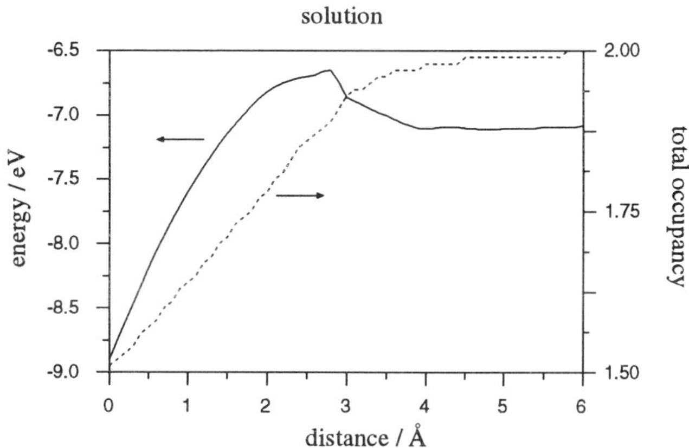

Figure 4. Energy and total occupancy for the iodide/iodine system near the surface of a Ag(111) electrode.

of the solvation sheath which occurs before the adsorption. The adsorbed state lies well below the ionic state in the solution, and carries a negative excess charge of $Q \approx 1/2$. In contrast to adsorption from the vacuum (see Fig. 1) the charge varies smoothly as the ion approaches the surface, and the non-magnetic state is favored throughout.

Conclusion

The formalism that we have used for the construction of potential-energy surfaces has several advantages: it is firmly based on quantum mechanics, and is computationally less demanding than other approaches since the computer simulations and the quantum-chemical calculations can be performed separately. However, it involves several approximations, which have been detailed above and in the cited literature [1, 2], and which limit the accuracy of the results. In particular, it has been assumed that the solvent is coupled linearly to the electron transfer. Although this seems to be a good approximation for some cases such as the $Fe^{2+/3+}$ reaction [9, 10], it need not be so in others. Still, we believe that these surfaces help in the understanding and the visualization of electrochemical reactions, and that they give the right order of magnitude for reaction barriers.

The main alternative to our approach are Car-Parinello type simulations of the interface such as have been reported in [11, 12] or similar quantum-chemical simulations [13]. These are potentially more powerful; at the present time their accuracy is limited by computing power, which necessitates a number of approximations, but this will improve with time. The two approaches may actually complement each other in the future: our method can give good approximations for a number of systems at comparatively little computational expense, while Car-Parinello simulations may give more exact results, but the long computations required will restrict their application to selected systems.

A major problem is the calculation of surfaces at other potentials than the point of zero charge. At present, it is not possible to simulate the electrochemical interface with a sufficiently high number of ions in order to include the space-charge region. Therefore the variation of the potential near the interface has to be introduced a posteriori, if it is included at all. This problem may, however, be relieved with further progress in computer technology.

References

[1] W. Schmickler, *Electrochim. Acta,* **41** (1996) 2329.

[2] W. Schmickler, *Chem. Phys. Lett.* **273** (1995) 152.

[3] P. W. Anderson, *Phys. Rev.* **124** (1961) 41.

[4] D. N. Newns, *Phys. Rev.* **178** (1969) 1123.

[5] V. L. Moruzzi, J. F. Janak, and A. R. Williams, *Calculated Electronic Properties of Metals*, Pergamon Press, New York, 1978.

[6] W. Schmickler *Interfacial Electrochemistry*, Oxford University Press, New York, 1996.

[7] M. T. M. Koper and R. A. van Santen, *Surf. Science* **422** (1999) 118.

[8] O. Pecina, W. Schmickler, and E. Spohr, *J. Electroanal. Chem.,* **394** (1995) 29.

[9] D. A. Rose and I. Benjamin, *J. Chem. Phys.* **100** (1994) 3545.

[10] J. B. Strauss, A. Calhoun, and G. A. Voth, *J. Chem. Phys.* **102** (1995) 529.

[11] J. W. Halley, *Electrochim. Acta* **41** (1996) 2229

[12] J. W. Halley, A. Mazzolo, Y. Zhou, and D. Price, *J. Electroanal. Chem.* **450** (1998) 273.

[13] A. Calhoun, M. T. M. Koper, and G. A. Voth, *J. Phys. Chem. B* **103** (1999) 3442.

Chapter 5

Modeling Metal Dissolution in Aqueous Electrolyte: Hartree–Fock and Molecular Dynamics Calculations

R. I. Eglitis[1], S. V. Izvekov[1], and M. R. Philpott[1,2,*]

[1]Institute of Materials Research and Engineering, 3 Research Link, Singapore 117602
[2]Department of Materials Science, National University of Singapore, 10 Kent Ridge Crescent, Singapore 119260

Aspects of the motion of atoms on metal surfaces during dissolution in aqueous electrolyte have been studied by two quantum computational methodologies. First INDO, the semi-empirical Hartree-Fock intermediate neglect of differential overlap method, was used to explore the potential energy surface of adsorbed water molecules and to speculate, using simple cluster models, on the reaction path followed by metal ions leaving the surface. Examples explore the potential energy curve for a metal atom leaving the surface and entering solvation cages of different geometry and show that the proximity of a local anion is crucial. Second a simplified Car-Parinello dynamics is used to examine dissolution in the time range 1 to 20ps. It is used to follow the development of the solvation shell around a departing ion and to calculate the potential of mean force acting on the ion.

Quantum and statistical treatments of the dissolution of metal in the presence of aqueous electrolyte is a challenging theoretical problem. In laboratory experiments the electrode is often maintained at a fixed potential which means that the surface charge of the electrode fluctuates. During dissolution the electrode has excess surface charge in the top most atomic layers. This charge is carried away by the flow of ions leaving the surface. The surface charge favors the adsorption of solvent and oppositely signed ions and consequently also the formation of surface complexes. Fluctuations in solvent adjacent to the metal can lead to configurations that stabilize the positive charge on surface atoms and so favor them breaking away from the surface and leaving as solvated cations. This paper briefly reports some of the work done to probe these complex problems. We concentrate on the dissolution of surface atoms or adatoms. Consideration of surface halide complexes like MX_2^- [1] is the subject of a separate study.

A two component strategy is used to describe the properties of the electronic potential energy surface (PES) of a metal atom (or ion) leaving a metal surface in presence of aqueous electrolyte. First we use a semi-empirical quantum electronic structure method to map out pieces of the potential energy surface involved in metal dissolution. The INDO method [2-4] provides a particularly convenient and fast method for calculating PE curves and surfaces and is applicable because the PES of interest is mostly outside the metal. The INDO version used has been previously parameterized for oxide, semi-conductor and ferroelectric materials [5,6]. Secondly, we probed the picosecond time scale dynamics of atomic nuclei and the metal electronic charge using a simplified version of an ab initio molecular dynamics method. The dynamics of dissolution were followed using a specialized MD scheme constructed in the spirit of the Car-Parrinello method by combining density functional (DF) calculation of metal cluster electronic functions and classical MD for metal ions and surrounding electrolyte solution. In order to restrict the number of electronic degrees of freedom the metal was modeled as a collection of Li atoms. The chemistry was restricted so that no chemical reactions occurred between Li and water.

Finally we conclude this introduction by noting that the dissolution of metal is an important step in many important technologies, including microelectronics [7], electro-crystallization [8], battery action [9,10], and the nano-restructuring of surfaces with local probe [11].

Summary of the INDO Method

In the study of the electronic structure and geometry of complex systems containing hundreds of atoms, there is a need to use methods that though not having the generality or flexibility of ab initio electronic structure methods are nevertheless accurate and fast enough for specialized systems through careful parameterization. We elected to use the INDO method, which is a semi-empirical version of *ab initio* Hartree-Fock formalism. The INDO method has parameters which are more or less transferable for the specific chemical constituent given, and is not subject to adjustment for each new compound. In recent years the INDO method has been used in the study of defects in bulk and on surfaces, e.g. in many oxide materials [3,4] as well as semiconductors. We recently applied this method to the study of phase transitions and frozen phonons in $KNbO_3$ [5], pure and Li-doped $KTaO_3$ [6] as well as F centers and hole polarons in $KNbO_3$ [12,13]. In the present work we used the set of INDO parameters for Li, O, H, Cl, Na that were used in Ref. [3-6,12,13].

The Fock matrix elements in the modified INDO approximation [3,4] contain a number of semi-empirical parameters. The orbital exponent ζ enters the radial part of Slater-type atomic orbitals:

$$R_{nl}(r) = (2\zeta)^{n+\frac{1}{2}} [(2n)!]^{-\frac{1}{2}} r^{n-1} \exp(-\zeta r) \qquad (1)$$

where n is the main quantum number of the valence shell. We used a valence basis set including 2s, 2p atomic orbitals (AO) for Li, 2s, 2p for O and 1s for the H atom. The diagonal matrix elements of the interaction of an electron occupying the μ th valence AO on atom A with its own core are taken as:

$$U^A_{\mu\mu} = -E^A_{neg(\mu)} - \sum_{v \in A} (P^{(o)A}_{vv} \gamma_{\mu v} - \frac{1}{2} P^{(o)A}_{vv} K_{\mu v}) \tag{2}$$

where $P^{(o)A}_{vv}$ are the diagonal elements (initial guess) of the density matrix, $\gamma_{\mu v}$ and $K_{\mu v}$ are one-centre Coulomb and exchange integrals, respectively. $E^A_{neg(\mu)}$ is the initial guess of the μ th AO energy. The matrix elements of an interaction of an electron on the μ th AO belonging to the atom A with the core of another atom B are calculated as follow:

$$V^B_\mu = Z_B \{ 1/R_{AB} + [< \mu\mu \,|\, vv > - 1/R_{AB}] \exp(-\alpha_{AB} R_{AB}) \} \tag{3}$$

where R_{AB} is the distance between atoms A and B, Z_B is the core charge of atom B. The adjustable parameter α_{AB} characterises the finite size of the atomic core B and the diffuseness of the μ th AO. The matrix element $< \mu\mu \,|\, vv >$ is the two-centre Coulomb integral. The resonance-integral parameter $\beta_{\mu v}$ enters the off-diagonal Fock matrix elements for the spin component u:

$$F^u_{\mu v} = \beta_{\mu v} S_{\mu v} - P^u_{\mu v} < \mu\mu \,|\, vv > \tag{4}$$

where the μ th and v th AO are centred at different atoms. Here u is an electron subsystem with α or β spin. The overlap matrix $S_{\mu v}$ is between electrons on the μ th and v th AO's. The product $P^u_{\mu v} < \mu\mu \,|\, vv >$ is the spin-density matrix.

The INDO parameterisation contains the following set of parameters per atom: the orbital exponent ζ describing the radial part of the μ th Slater-type AO on the A-atom; the electronegativity of the A-atom $E^A_{neg(\mu)}$, defining the μ th AO energy; the μ th AO population $P^{(0)A}_{vv}$, i.e. the 'initial guess' for the diagonal element of the density matrix; the resonance integral $\beta_{\mu v}$ entering the off-diagonal Fock matrix element where the μ th and v th AOs are centred at the different atoms A and B; and the adjustable exponent α_{AB} was as previously described.

Metal dissolution potential energy curves from INDO calculations

Some of the key issues in metal dissolution are associated with water molecules binding to the metal surface, the shape and position of solvent cages able to stabilise the metal cation , the course of the back flow of charge from the departing ion to the metal . These are complex issues that can be treated only approximately. In this section we report briefly on results for water adsorption on (100) surface, and the stabilisation of the metal cations by hydration cages with given geometry (tetrahedron, square base pyramid), and local fields from anions.

The inset of Figure 1 shows the cluster model for a single water molecule siting on the (100) surface of a metal. The cluster consists of 63 atoms of Li. The water molecule was located on the central top site, or on the adjacent bridge site or on a hollow site. Curves labelled 1, 2, and 3 refer to top, bridge or hollow on the (100) surface. The oxygen atom of the water molecule points at the adsorption site, and H atoms point away. According to the calculated PE curves shown in the main figure, the top site binding energy of a single water molecule is approximately 0.7 eV, when both hydrogen atoms are pointing away from the surface. The energy minimum occurs when oxygen atom is 3.69au from the (100) surface plane. Tilting the water molecule 70° gives an additional energy gain of approximately 0.1 eV. The binding energy of the bridge site is 0.41 eV at 3.9 au from the (100) surface, and the hollow site has a binding energy of 0.28 eV at 3.95 au. These results are consistent with previous calculations for the binding of water to clusters of metal atoms [14-20] and surfaces [19-22]. Of course these calculations do not include relativistic effects which are important for third row transition elements [23] and so only apply to low Z materials.

Next we describe calculations aimed at exploring the role of solvent cage shape and size in stabilising a charged surface atom, as well as the effect of nearby halide species. First we calculated the binding energy of the centrally positioned Li surface atom (not an adatom) at different distances from its position in a 63 metal atom cluster in a vacuum (see curve 1 in Figure 2). In these calculations the middle Li atom from (100) surface was moved in [100] direction. INDO calculations showed that the binding energy was approximately 2.13eV. Next a cage of four water molecules with oxygen atoms centred at the corners of tetrahedron was placed face down (three waters parallel to (001)) on the surface. Then the PE curve for moving the central surface atom into the cage was calculated. See curve 2 and inset in Figure 2. Finally a chloride ion was placed 15.40au above the (100) surface and a sodium ion was placed 15.40au behind the cluster, both ions on the cluster symmetry axis. The chloride ion simulates the presence of halide ion near the surface and the sodium ion insures charge neutrality for the entire system. The PE curve for moving the central surface atom was then recalculated to give PE curve 3 in Figure 2. Note how the barrier drops enough for a Li atom to depart as a cation stabilised by the water cage and the presence of Cl⁻ ion. This was the key result of the INDO calculations, namely that the water and ions were both needed for metal dissolution in the cluster model. This rather

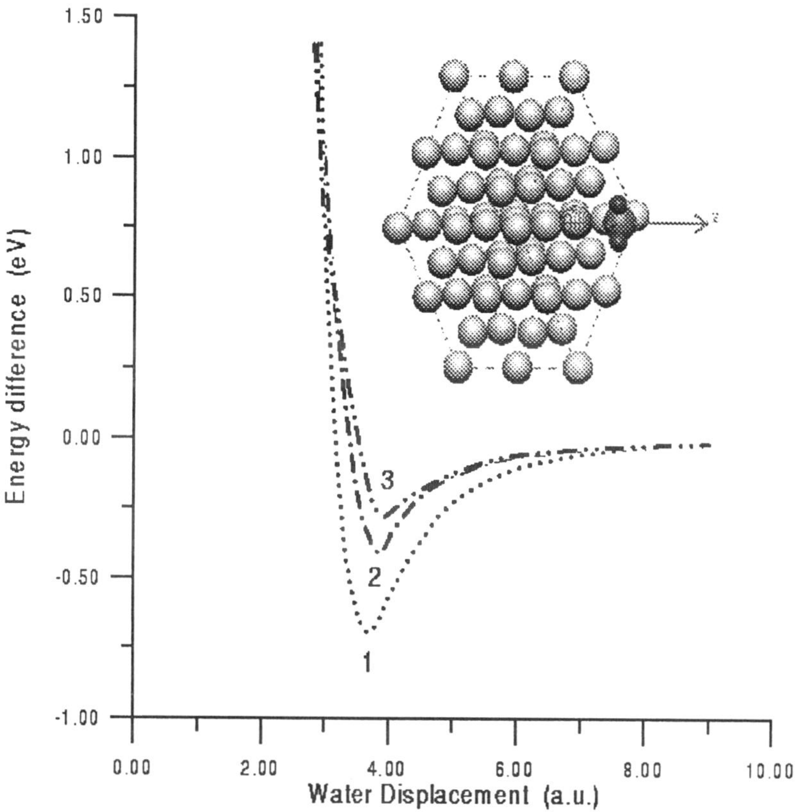

Figure 1. Binding energy curves for water molecules on top site (curve 1), bridge site (curve 2) and hollow site (curve 3) on the (100) face of an M_{63} cluster of metal atoms. All calculations performed using INDO.

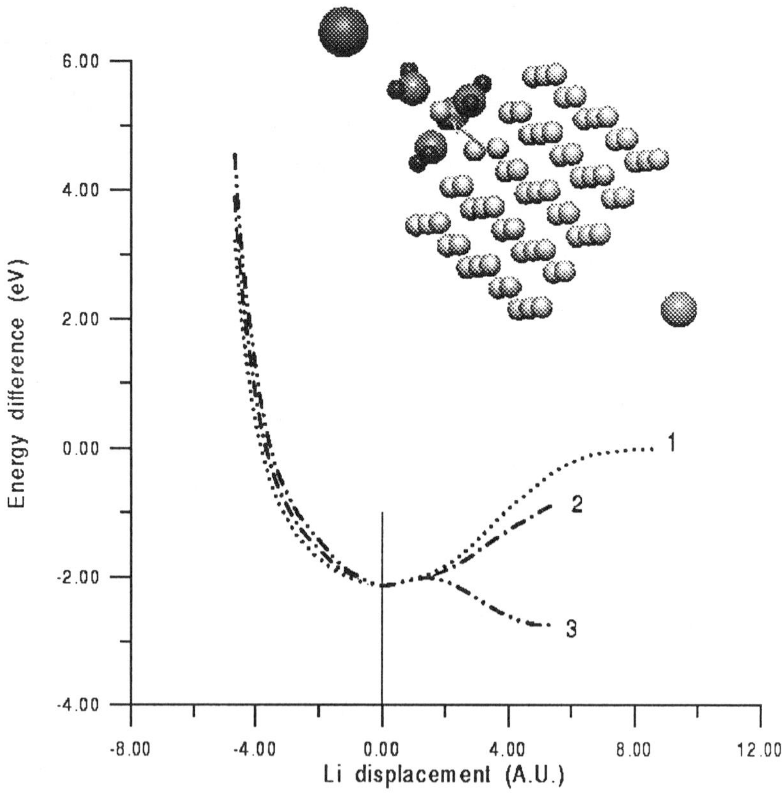

Figure 2. Potential energy curve for 63 atom Li cluster. For each case the central surface Li atom is moved along z. Curve 1 is for the cluster in vacuum. Curve 2 is for case where a tetrahedral water cage is located above the lattice atom being moved and the shift is due to water stabilization alone. Curve 3 is the potential energy curve when a chloride ion Cl^- is located on the z axis above the central atom site at a distance of 15.40 au Simultaneously a Na^+ ion is located below the cluster on negative z axis.

unsophisticated calculation does not include important solvent effects like the effect of dielectric constant or screening of the surface charge by a dynamic Helmholtz electric double layer. Nevertheless it does illustrate how stabilisation of the departing atom as a cation is the result of several effects. That this result is not peculiar for either a surface atom or a tetrahedral solvent cage is shown by the set of PE calculations summarized in Figure 3.

In Figure 3, we considered an adatom on a four hold hollow site of the (100) surface. For this case we calculated the PE curve as the Li adatom left a hollow site on a 55 Li atom cluster (001) surface. In the inset of Figure 3 is a cartoon of the 55 atom cluster consisting of one adatom sitting on a 54 atom cluster. As the first step we calculated the path for Li adatom leaving a 55 atom cluster. Our calculations showed, that the Li adatom extraction path started with a gradual energy lowering of system until the adatom reached an energy minimum positioned at 3.19 au from the cluster (001) surface. The energy minimum of Li adatom occurred at -1.27eV. According to these calculations the Li adatom dissociation energy, namely energy difference between Li adatom at its energy minimum position (-1.27 eV) and at 10.15 au from cluster (001) surface (-0.356 eV) was 0.914 eV.

To study the effect of solvent cage shape we replaced the tetrahedron with a square based pyramid of five water molecules above the Li (100) surface. The first layer, containing four water molecules in a square plane was located 3.69 au above the top of Li (100) surface. Since there were repulsions among four oxygen molecules in the plane parallel to the Li (100) surface, the four oxygen atoms were not on top sites, but displaced by 8% of the Li crystal lattice constant in the directions (100) and (010). See the inset model in Figure 3. Hydrogen atoms of four planar water molecules were positioned so as to maximise the distance between hydrogen atoms of different water molecules. Planes containing hydrogen atoms of four planar molecules were tilted by angle 70 degrees from normal to (100) Li surface. Oxygen atom of fifth apical water molecule was located 8.79au above the remaining Li cluster (100) surface.

The next calculation involved moving the Li adatom away from the 55 atom Li cluster into the cage of five water molecules (refer to Fig. 3, curve 2). In the presence of the five water molecules the first stage of removal of the Li adatom reached an energy minimum at 3.404 au from (100) surface. This minimum was at a relative energy -1.296eV. This showed that moving further from (100) surface in the presence of the five water molecules was an energetically unfavourable process, eventually reaching a value of -0.95eV somewhere around 6.939au from the (100) surface plane. The general conclusion for the case of Li adatom movement away from the 55 atom cluster via cage of five water molecules (see curve 2, Figure 3) is that the 0.346eV high energy barrier prevents cluster dissolution.

Finally in the last calculation with this model, we added a NaCl ion pair to this system. The Cl⁻ ion (with effective charge -1e) was located 15.40au above the top of Li cluster (100) surface and the Na ion (with effective charge +1e) located 15.40 au below the 55 atom Li cluster. In the first step of Li adatom extraction the Li adatom reaches an energy minimum at 3.8au from Li cluster (100) surface (PE value

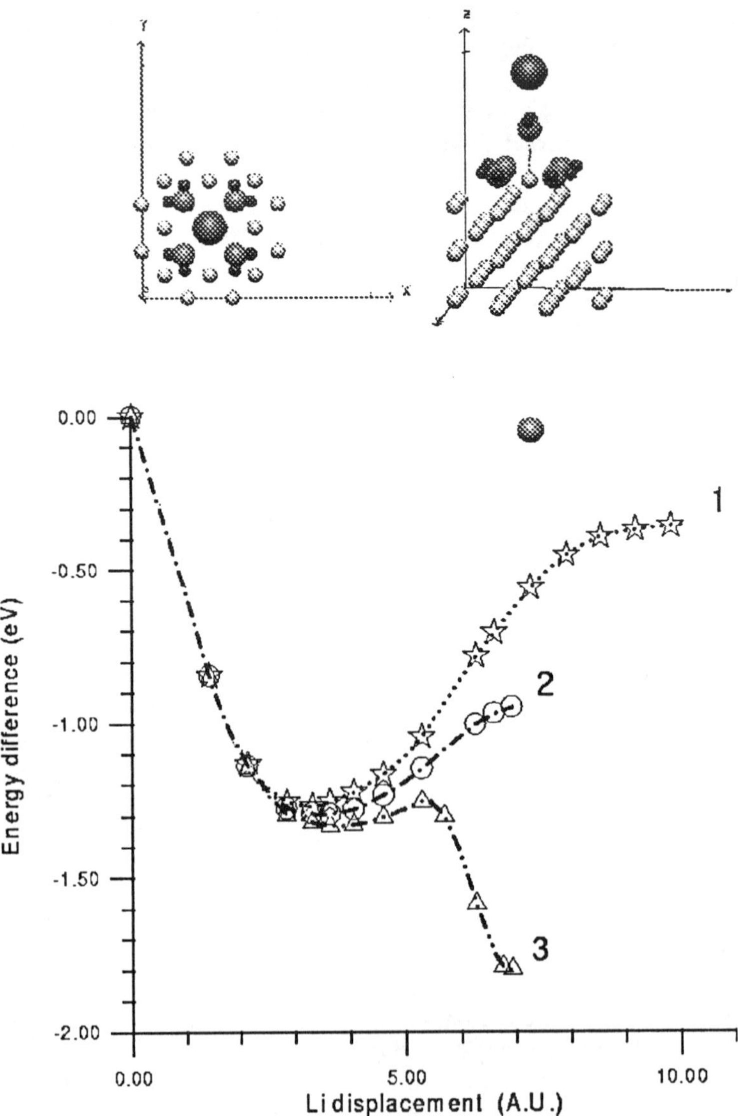

Figure 3. The PE curves for an adatom in a 4H-(100) site on a 55-atom Li cluster with the adatom moved along z. Curve 1 is for a cluster in vacuum, a neutral adatom is removed. Curve 2 is for case where a cage containing five water molecules is located above the adatom site, and the shift is due to water stabilization. Curve 3 is the potential energy curve when a chloride ion Cl⁻ is located on the z-axis above the central atom site at a distance of 15.40au. Simultaneously a Na⁺ ion is located below the cluster on negative z axis.

-1.33eV). At larger distances the adatom energy increases a little to a maximum (-1.254 eV) at 5.28au from Li (100) surface. When Li adatom is completely inside cage of five waters, and is in the presence of Cl^- and Na^+ atoms on the top and bottom of system, then there is a large energy lowering. At this point Li adatom is a positive ion, the energy reaches its minimum value (-1.799eV) at 6.85au from Li cluster (100). These calculations show that Li adatom dissolution via square pyramid containing five water molecules in presence of Cl^- and Na^+ ions is energetically favourable process. The energy gain due to the dissolution process was 0.545eV.

These INDO calculations demonstrate that a solvent cage and an electric field from a local anion are needed for dissolution to occur.

Ab initio molecular dynamics simulation

In a completely different approach that emphasizes motion we have performed a series of molecular dynamics (MD) simulations of a metal cluster immersed in aqueous sodium chloride solution. In our approach we combined an *ab initio* quantum mechanical treatment of the electrons of the metal cluster with classical mechanics simulation of all the nuclei in the system. This includes not only the metal ions of the cluster but also all the nuclei making up the aqueous electrolyte phase (water and ionized sodium chloride). The basic problem was to simultaneously solve for the electronic and ionic motion in the metal in the presence of a liquid electrolyte phase. First we describe the model used. All the electrons of the metal cluster were confined to a cubical box of edge length L_{el} (to be called the "electronic box"). The electronic wave functions were set to zero on the box sides and outside of the box. A sinusoidal wave basis was used, which satisfies the condition of zero charge on the walls of the electronic box. The Kohn-Sham orbitals in such a formalism were real functions. The introduction of zero charge density boundary condition was reasonable because the metal ground state charge density rapidly dropped to zero with distance from the cluster. The separation of the cluster from the electronic walls was crucial. If it was too small then many of the calculated basis functions were compromised. In this calculation scheme to achieve good accuracy a box size must be chosen to ensure that the walls occur in the region where valence charge density is close to zero. The electronic box in this formalism should be considered as formed from walls with infinitely high potential trapping all the electrons inside. *All chemistry is confined to this box.*

The entire system (cluster and electrolyte) was enclosed in a much larger cubical box (called the "molecular dynamics" or MD box) with edge length L_{MD}. The interaction of the electrolyte constituents with the MD box was was subject to some experimentation since the idea was to reduce wall effects to a minimum. We followed Car-Parrinello scheme [24] in local density (LD) and Bachelet-Hamann-Schluter (BHS) pseudopotential approximations [25,26] to simulate the dynamics of the electronic charge of the metal cluster. To extend the MD simulation to larger time scales the Car-Parrinello dynamics for the valence electronic functions was performed in the basis of lowest energy Kohn-Sham orbitals calculated for a given

60

reference geometry. This geometry was chosen to be that of an ideal cluster in vacuum. The full set of Kohn-Sham orbitals is complete. It turns out in that the Kohn-Sham orbitals for a geometry different (in some small extent) from the reference one could be well approximated with linear combinations of a sufficiently large number of the lowest energy Kohn-Sham orbitals of the reference geometry. For small perturbations the number of restricted Kohn-Sham orbitals in the basis could be kept small, and still produce good accuracy in the calculation of the cluster electronic structure. The accuracy of this representation was checked by independent DF calculations in the original plane wave basis.

Space does not permit more than a cursory statement of the equations underlying Car-Parinello dynamics see reference [24]. The one electronic valence wave functions were expanded into linear combinations of the N lowest energy Kohn-Sham orbitals of the reference geometry of the cluster in vacuum as

$$\psi_i(r) = \sum_{j=1}^{N} c_i^j \psi_j^o(r) \tag{5}$$

The equations of motion are

$$\mu \ddot{c}_i^j = -\frac{\partial E_{tot}}{\partial c_i^j} - \sum_{j=1}^{N} \lambda_{ij} \frac{\partial \sigma_{ij}}{\partial c_i^j} \tag{6}$$

for the electronic degrees of freedom and

$$M_I \ddot{R}_I = -\nabla_I E_{tot} \tag{7}$$

for all the nuclear degrees of freedom. In equations (6) and (7) the symbol μ is the fictitious electronic mass, $\{ M_I \}$ are the masses of the cations or atoms, and the set of parameters $\{\lambda\}$ are the Lagrangian multipliers chosen to satisfy the following orthonormality constrained equation

$$\sigma_{ij} = \sum_{m} c_i^m c_j^m - \delta_{ij} = 0 \tag{8}$$

Energy E_{tot} includes the interaction of ionic and electronic subsystems of the cluster with electrolyte surroundings. Note that the unperturbed basis $\{\psi_j^o\}$ is only calculated once by conventional DF method before starting the Car-Parrinello dynamics. The accuracy in the representation of one-electron wave functions is controlled by the number N of lowest Kohn-Sham orbitals included in the basis. Our work suggests that for systems like the ones studied here, the size of the basis can be chosen to be rather small (several times larger than a number of occupied electronic states) and still achieve reasonable accuracy in DF calculations. Therefore this scheme was much faster than some previously developed methods. There have been several methods that provide a numerical scheme to treat simultaneously the electronic structure of metals using DF formalism together with classical motion of adjacent or surrounding electrolyte for large time scales. In particular J.W. Halley and co-workers proposed [27] a steepest descent simple iteration technique to adjust the one electron wave functions to the changing electrolyte configuration. The advantage of our scheme is small computational effort for the simulation of metal electronic structure at fairly good accuracy. This is because we operate only with one-electron amplitudes in a relatively small-restricted Kohn-Sham basis, calculated for the initial metal reference geometry.

For computational simplicity the atoms were taken to be Li atoms (*sp*-bonded metal). The cluster consisted of 35 one-electron Li atoms surrounded by aqueous electrolyte described by the SPC/E model [28] for water and simple ions (NaCl). Electrolyte was 1.65 molal solute of NaCl and consisted of 800 water molecules and 48 ions. In the ab initio MD part Car-Parrinello dynamics of only one cluster surface ion was considered, with all other metal cluster ions on fixed sites. This ion was chosen to be central ion on (001) cluster surface. When the free metal ion left the electronic box it become a classical object like the other simple ions in the electrolyte. To speed up the calculations for the moving ion the projected part of BHS pseudopotential was tabulated on a fine grid along normal direction to a cluster (001) surface. This Li^+ ion was only allowed to move in the z direction. The integrals corresponding to the Hartree part of the total electronic energy of the cluster were evaluated by fast Fourier transform techniques.

A difficult question to decide was how to select a good short-ranged potential (non-coulomb part) for water-lithium ion interactions for the Li cations in the solvent and those belonging to the metal cluster. We took this short-ranged interaction to be of the Lennard-Jones (6-12) type. The interaction of the water and Na^+Cl^- ions with cluster electron density was another difficult issue. We chose this interaction to be composed of two parts. The first purely electrostatic part was taken to be the same as produced by the charge distribution of the SPC/E water model. The second term was taken to be a spherically symmetric polarization term, centred on the oxygen nucleus. This model did not explicitly include a term for Pauli repulsion in the metal charge density - electrolyte interaction. However the potential seems to be good enough to reproduce the development of the solvation shell as Li leaves the cluster and enters the electrolyte phase.

The coupled sets of equations which represent the Car-Parrinello dynamics of the cluster structure (Eqs. (6), (7)) and the classical MD of the aqueous solution were integrated simultaneously. The number N of lowest energy Kohn-Sham orbitals in basis set for Car-Parrinello simulation (see Eq. (5)) was chosen to be twice the number of occupied states. This basis produced good accuracy in the representation of the cluster electronic structure; the error was less than 5% as compared with exact DF calculations. The fictitious electronic mass was 10,000au, which corresponds to a relaxation time of 20fs for the charge density during the oscillation of the surface ion.

The frequency of the oscillation was approximately 92fs (365cm^{-1}). The binding energy of the cluster per atom was 1.2eV. The binding energy of the central atom on (001) cluster surface was 2.11eV. This was in agreement with INDO calculations presented above. The binding energy for the ion was 7.33eV, so that the affinity for the cluster was around 0.2eV. This is reasonable because for the size of the cluster we studied, the affinity has to be significantly smaller than work function. We have found independent LDA calculations to verify the affinity for our case. The binding energy for the same ion obtained from *ab initio* MD was 7.4eV.

We studied the influence of the electrolyte on several cluster properties. For example the binding energy of the ion with the metal cluster and the charge distribution inside the cluster. We also calculated potential of mean force (PMF) that governs the dissolution process. The other part of our simulation was concerned with the structure of the solvation shell of metal ion during the "dissolution" event. The formation of solvation shell correlates well with shape of PMF curve. These calculations required us to collect statistical data for periods of several tens of picoseconds. Therefore being able to perform relatively long time simulations proved to be one definite advantage of the methodology described here.

The computed PMF curve exhibited a well-defined maximum at 3.5-4.0au from the cluster surface. The barrier, which separated ion in the cluster and in electrolyte, was rather high at about 1.80eV. This means that the neutral cluster would not dissolve at room temperature when surrounded by the small pocket of electrolyte as confined by the MD box of the size actually used. This could be remedied in three ways. First with a larger embedding electrolyte cluster so that more of the full dielectric response of the electrolyte could develop. Second by applying an external electric field to mimic potentiostatic tuning of the electrode. Third by repeating the calculations with charged clusters.

Finally we address the question of how the solvation shell develops as the ion leaves the metal. A specific result is shown in Figure 4. The radial distribution function $g(r)$ and the hydration number $n(r)$ (integral of $g(r)$) were calculated for an ion fixed at different distances d = 0.0, 4.0 and 8.0au from the cluster surface. The first peak of the radial distribution function formed in 5-10ps of simulation time. When the Li ion was in the region of bulk solvent (d = 8au) the first solvent shell contained five water molecules. In a previous classical MD simulation Lee and Rasiah calculated the solvation number for Li$^+$ ion in SPC/E water to be about 4.1 [29]. The first hydration shell first becomes complete, i.e. formed from four molecules at a distance of approximately 4 -5au. This also corresponded to the position where the first maximum is the PMF curve peak occurred. Note too that the radial distribution function has a fully developed second peak (about 7.5au in Fig. 4 bottom panel) only when the ion was about 8au from the surface.

Conclusions

The combination of semi-empirical INDO method and an *ab initio* MD method provides complementary insights into metal dissolution in aqueous electrolyte. INDO

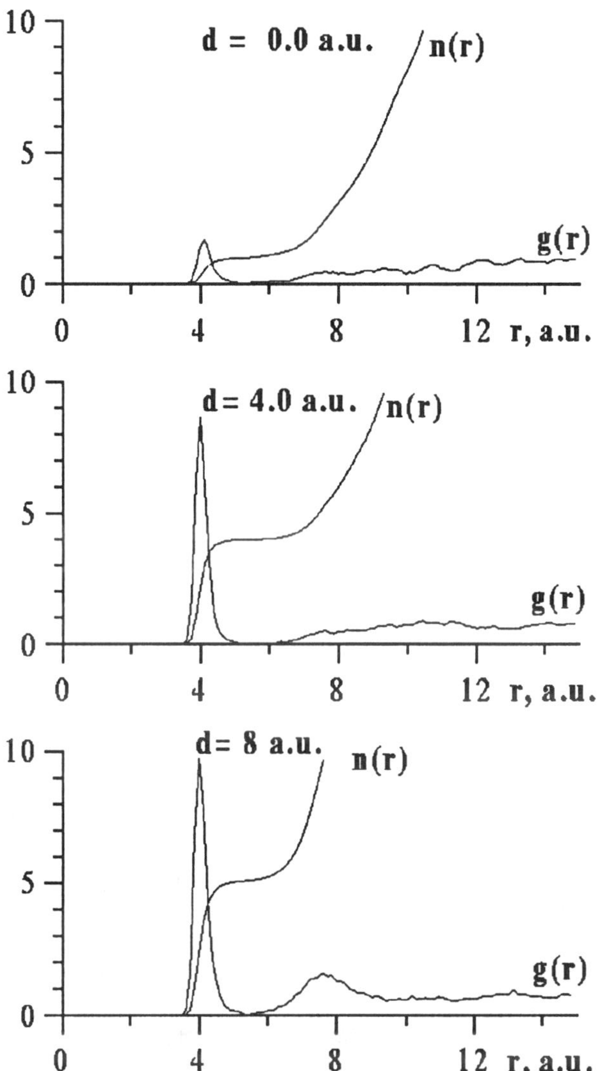

Figure 4. Change in hydration shell of metal ion with distance from surface. Radial distribution function g(r) and running integration number n(r) for Li ion separated by a distance d from the cluster. Top, d = 0.0au, the metal ion in the (100) surface plane has one water in its solvation shell. Middle, d = 4.0au, metal ion has four waters in the first shell. Bottom, d = 8.0au the ion has five waters in the first solvation shell.

studies showed that a water hydration cage and anion were needed for dissolution. The MD studies revealed that formation of the hydration cell was a more complicated process than would have been envisaged by the INDO model.   The evolution of the solvation shell correlated with features in the calculated potential of mean force (PMF) curve of the dissolution event. The first solvation shell progressively added water molecules from $n= 1$ at d = 0.0 au to $n = 5$ at d = 8 au. The first hydration shell was already completed at the distance where a peak in the PMF curve occurred. At this point the metal ion appeared fully hydrated because the appearance of a second well defined peak at 7.5 au in the radial distribution function corresponding to the formation of the outer part of the hydration sphere.

References

1. O.M. Magnussen and R.J. Behm, "Atomic-scale processes in Cu Corrosion and Corrosion Inhibition" , Mat. Res. Soc. Bulletin, July 1999, pp. 16 - 23.
2. Pople J.A.; Beveridge D.L. *Approximate Molecular Orbital Theory*, McGraw-Hill, New York, 1980.
3. Stefanovich E.; Shidlovskaya E.; Shluger A.; Zakharov M. *Phys. Stat. Sol. (b)* 1990, Vol. 160, pp. 529 - 536.
4. Shluger A.; Stefanovich E. *Phys. Rev. B* 1990, Vol. 42, pp. 9664 - 9673.
5. Eglitis R.I.; Postnikov A.V.; Borstel G. *Phys. Rev. B* 1996, Vol. 54, pp. 2421 - 2427.
6. Eglitis R.I.; Postnikov A.V.; Borstel G. *Phys. Rev. B* 1997, Vol. 55, pp. 12976-12981.
7. Chang C.Y.; Sze S.M., *ULSI Technology*, McGraw-Hill, New York, 1998.
8. Budevski E.; Staikov G.; Lorentz W.J., *Electrochemical Phase Formation and Growth*, VCH, Publishers, NY, 1996.
9. Wolverton C.; Zunger A. *Phys. Rev. Lett.* 1998, Vol. 81, pp. 606-609.
10. Ceder G.; Chiang Y.-M.; Sadoway D.R.; Aydinol M.K.; Jang Y.-I.; Huang B. *Nature* 1998, Vol. 392, pp. 694 - 696.
11. Kolb D.M.; Ulmann R.; Ziegler J.C. *Electrochimica Acta* 1998, Vol. 43, pp. 2751 - 2760.
12. Eglitis R.I.; Christensen N.E.; Kotomin E.A.; Postnikov A.V.; Borstel G. *Phys. Rev. B* 1997, Vol. 56, 8599-8604.
13. Kotomin E.A.; Eglitis R.I.; Postnikov A.V.; Borstel G.; Christensen N.E. *Phys. Rev. B* 1999, Vol. 60, pp. 1 - 5.
14. Holloway S.; Bennemann K.H. *Surf. Sci.* 1980, Vol. 101, pp. 327-331.
15. Head J. *Surf. Science* 1994, Vol. 318, pp. 204-207.
16. Yang H.; Whitten J.L. *Surf. Science* 1989, Vol. 223, pp. 131 - 150.
17. Ribarsky M.W.; Luedtke W.D.; Landman U. *Phys. Rev. B*, 1985, Vol. 32, pp. 1430 – 1439.

18. Bauschlicher C.W. *J. Chem. Phys.* 1985, Vol. 83, pp. 3129 – 3136.

19. Siepmann J.I.; Sprik M. *J. Chem. Phys.* 1995, Vol. 102, p. 511 - 524.

20. Zhu S.-B.; Philpott M.R. *J. Chem. Phys.* 1994, Vol. 100, pp. 6961 - 6968.

21. Foster K.; Raghavan K.; Berkowitz M. *Chem. Phys. Letters* 1989, Vol. 162, pp. 32 - 38.

22. Spohr E.; Heinzinger K. *Chem. Phys. Lett.* 1986, Vol. 123, pp. 218 – 222.

23. Pacchioni G., *Electrochimica Acta* 1996, Vol. 41, pp. 2285 - 2291.

24. Car R.; Parrinello M. *Phys. Rev. Lett.* 1985, Vol. 55, pp. 2471 - 2474.

25. Kohn W.; Sham L.J., *Phys. Rev.* 1965, Vol. 140, pp. A1133 – A1141.

26. Bachelet G.B.; Hamann D.R., *Phys. Rev. B* 1982, Vol. 26, pp. 4199 – 4228.

27. Price D.; Halley J.W., *J. Chem. Phys.* 1995, Vol. 102, pp. 6603-6612.

28. Berendsen H.J.C.; Grigera J.R.; Straatsma T.P. *J. Phys. Chem.* 1987, Vol. 91, pp. 6269 – 6271.

29. Lee S.H.; Rasaih J.C, *J. Phys. Chem.* 1996, Vol. 100, pp. 1420 - 1425.

Chapter 6

Electron Wave Packet Propagation Studies of Electron Transfer at the Semiconductor–Liquid Interface

B. B. Smith

National Renewable Energy Laboratory, 1617 Cole Boulevard, Golden, CO 80401–3393

This paper examines the implications of three particular techniques for simulations of time dependent electron transfer (ET) at semiconductor-liquid interfaces (SLI's). The model examined uses a one-electron method employing wave packets, pseudopotentials, and molecular dynamics, an approach that we dub wave packet molecular dynamics (WPMD). We analyze details of the mechanism for SLI ET by using the methodology. The model is versatile enough to address conventional SLI ET, surface state and adsorption mediated ET, photoexcited ET, and ET between quantum dots and other microstructures. The most sophisticated of the simulation techniques employed here is the surface hopping trajectory formalism.

The semiconductor-liquid interface (SLI) presents many theoretical challenges. Regarding its electronic structure, high level methods are too computationally demanding to address large spatial scales, and lower level methods can be highly inaccurate. Excited state properties remain difficult to address via any technique. Presently it is only through a comparison of many theoretical techniques that confidence in results can be gained, and ultimately experiment is the arbiter.

We have employed a wide range of electronic techniques ranging from tight binding, to semiempirical CNDO, INDO, and MNDO methods, density functional theory in many forms, as well as standard ab initio quantum chemistry. Our main goal has been to interface these electronic methods with molecular dynamics simulations to address, at various levels of sophistication, the coupling between the SLI nuclear and electronic dynamics.

Below we report on one of our many dynamic SLI simulation methods. This approach uses a model of the SLI system employing a single electron wave packet, pseudopotentials, and various molecular dynamics methodologies. We dub this general approach wave packet molecular dynamics (WPMD). Compared to most of our other techniques it treats the electronic subsystem very simplistically, i.e., it is a one electron approach which treats only the electron involved in the electron transfer quantum mechanically. However, as will be shown, it offers a number of insights into

SLI electron transfer (ET), and in some of its forms can address nonadiabatic processes. The formalism discussed here can be applied to essentially all the new systems of interest in the SLI ET community, especially the new time dependent experimental measurements (*1*).

Two aspects of WPMD are especially important (i) it is a tool to investigate the shortcomings of higher level smaller spatial scale simulations and (ii) it is a simulation method in its own right for investigation of SLI physics. We only have space to address the second point.

The WPMD Models and Methodology

The total wavefunction for the electronic and nuclear subsystems of any ET system satisfies the time dependent Schrödinger equation. But solving for the full coupled system of vibrational, rotational, and translational states is computationally intractable even for relatively small systems. The usual practice is to assume the nuclear system can be treated classically, and that only a few electronic degrees of freedom require quantum treatment. In this approximation the quantum and classical systems are still connected because changes in electronic distribution affect the nuclear forces, and in turn nuclear motion affects this distribution. Thus a self-consistent solution is necessary for detailed study. However, one can examine approximately how the nuclear trajectory influences the probability of transition between quantum states by assuming a fixed classical nuclear trajectory. The latter assumption is characteristic of conventional ET theories. We will use this assumption in most of the simulations reported herein, but some simulations we report below treat the nuclear and electronic motion in a self-consistent fashion, and we will compare the results.

Our WPMD simulations, which we have described elsewhere (*2*), are fully dynamic 3D simulations addressing the ET electron, solvent, redox species, and semiconductor. The nuclei are treated classically and only one electron is explicitly treated, that involved in the ET. This electron moves in a pseudopotential consisting of a superposition of atomic pseudopotentials. That is:

$$V_{system} = \sum_i V_i \tag{1}$$

where the V_i are the atomic pseudopotentials. Constructions like eq. 1 are typically termed empirical pseudopotentials (*3*), where the V_i are adjusted to fit experimental and ab initio data. The electron is treated quantum mechanically by using approximate solutions of the time dependent Schrödinger equation, which employs eq. 1 for the potential. The V_i of eq. 1 determine the electronic structure of the system; there are additional pseudopotentials used in some of our methods that determine the interatomic forces. The pseudopotentials used are described further below and in ref. (*2*). All methods use a single redox species in the simulation box, and a cluster of 300-1000 solvent molecules (depending on the rigor of the simulation method) and another cluster of 200-2000 semiconductor atoms. The boundary conditions used and related simulation details have been described elsewhere (*2*). The several different versions of WPMD that we use are described next.

The first and simplest WPMD method we employ propagates the electron wave packet in a fixed transition state (TS) nuclear configuration. We find this TS configuration using standard constrained classical molecular dynamics. (We define TS configurations as those where the total energy of the system is the same in presence of the reduced or oxidized redox species, and where (as discussed later) the redox species eigenstate involved in the ET is in the conduction band.) Then the electron wave packet can be propagated in that fixed potential via solution of the time dependent Schrödinger equation:

$$i\,\hbar\,\partial\,|\,\psi(\mathbf{r},t)\rangle\,/\,\partial t\,=\,H'\,|\,\psi(\mathbf{r},t)\rangle \tag{2}$$

for this system using numerical techniques discussed in ref.(2). Our electronic Hamiltonian is

$$H' = (-\hbar^2/(2\,m_e))\,\nabla^2 + V_{system} \tag{3}$$

with V_{system} given by eq. 1. We call this mode of our WPMD the frozen solvent regime. In this method the dynamics of the electronic and nuclear subsystems are obviously unconnected, since there is by definition no solvent dynamics.

We have various techniques to examine effects of solvent dynamics with our WPMD. One of these constitutes our second type of WPMD, and in it the nuclear trajectory is a fixed function of time. We call this fixed trajectory WPMD (FT-WPMD). The fixed nuclear trajectory is obtained from a fully classical MD simulation in ways described shortly. The important point is that in these fixed trajectory simulations the electronic subsystem (our ET electron density) does experience the changing potential that the dynamic polar solvent nuclei produce and this, of course, can cause a redistribution of electron density between the redox species and semiconductor. But the nuclear dynamics is fixed and classical and unconnected to the electronic dynamics, and its associated changing charge distribution, except in the following way. When electronic transitions are made in the simulation as a result of the changing potential, and total system energy is thus not conserved, we force energy conservation in an ad hoc fashion by changing the solvent and redox species atomic velocities at random. The latter is the same as the most primitive of ad hoc energy conservation techniques used in surface hopping trajectory methods (4). Note that the "fixed" nuclear trajectory actually does change as a result of the random redistribution of energy. But the overall perturbation of the trajectory can be small since energy conservation is accomplished by changing the velocity of each of the many solvent molecules, and by random amounts. Thus, while the qualitative character of the trajectory can indeed change, we search for and employ regimes where it remains essentially unchanged (same amount of charge transfer and transition state lifetime, as well as other quantities related to the ET) when different random number generator initialization seeds are used.

In the fixed trajectory simulations the nuclei are propagated in time steps of Δt, and the electron probability density is propagated for the same time (in time steps of $\Delta t/N$) by solving the time dependent Schrödinger equation (eq. 1-3) numerically (2), and where the potential is now time dependent due to solvent dynamics.

The manner in which we produce the nuclear trajectories for the fixed trajectory method warrants further explanation. We first do a constrained fully classical MD simulation that generates the TS configurations defined above, by using appropriate pseudopotentials. We take a sample of these configurations and use each of these for the initial configuration of unconstrained purely classical MD simulations that run the nuclear dynamics forward and backward in time relative to the initial transition state configuration. The outcome of this is that the trajectory is effectively run from the equilibrium dynamics of the reactant in the distant past to the far future relative to the time the system is in the TS.

The TS is a state of high energy, so it is relatively short lived. Thus if we start our fully classical simulation in the TS, and we do not change the charge on the redox species, the solvent will relax to configurations that are in equilibrium with the reactant. To "produce" the ET product in the fully classical simulation we must change the charge on the redox species by unit charge (single ET events are assumed). We change it instantaneously while it is in a TS configuration, and energy is thus conserved.

A third mode of our WPMD involves a method that addresses self-consistency between the nuclear and electron dynamics via the surface hopping trajectory (SHT) methodology. Many SHT methods exist, which have greater and lesser degrees of physical justification (4). In our SHT WPMD we currently employ two extensively utilized and reviewed methods. The first method is essentially Tully's fewest switches SHT method (5), which requires an ad hoc redistribution of energy after electronic transitions, as in our fixed trajectory method. Unlike our fixed trajectory method, this technique appropriately coherently propagates the dynamical mixture of states for all times(4); but its ad hoc energy redistribution clearly shows it is not fully self-consistent. We also employ a method suggested by Coker (4) which is a hybrid of the SHT method of Webster et al.(6, 7) and Tully's methods. This hybrid method does not require an ad hoc redistribution of energy since the energy is conserved automatically by iterating the trajectory to self-consistency and it uses Tully's method of coherent propagation over all time (rather than over a single time step as in the Webster et al. method (4, 6, 7)) . Numerous analogous self-consistent simulations employing classical and quantum subsystems exist (4, 8-13). The ultimate check on such approximate simulations are full quantum mechanical simulations. That is partly why we have pursued fully quantum mechanical MO-MD and Car-Parrinello simulation methods (13, 14). We explain the SHT techniques we use in some detail below.

We pause to note that Tully originally formulated the SHT method for charge transfer associated with atoms/molecules scattered off UHV solid surfaces. The SHT simulation method is thus a natural choice to employ for the related SLI ET problem, as we have discussed elsewhere (2).

The pseudopotentials employed in our WPMD in this paper are discussed in detail in ref. (2). They model a water solvent, a GaAs semiconductor, and redox species consisting of a transition metal with six octahedral water ligands. The redox species is essentially the ferrous-ferric system, which we employed for ease of comparison with homogeneous and metal interface studies. We used essentially all the same pseudopotential parameters for the redox species used in these numerous other ferrous-ferric studies (14-17), except we modified, in effect, the ionization potential to position the redox species ET state at a desired energy relative to the GaAs conduction band edge. A second redox species parameterization we used was exactly the same as that just described except a different electron-ligand (water ligand) pseudopotential was designed so that it trapped electron density in a more delocalized fashion than the ferrous/ferric system. The latter allowed us to examine wave packet flow into and out of the ligands. We used a local pseudopotential to represent (100) GaAs.

With this background we now introduce results from our SLI ET WPMD simulations.

Results

SLI Oxidation

We now investigate conduction band SLI oxidation (valence band processes are not addressed in this paper) with the frozen solvent and fixed nuclear trajectory WPMD SLI simulations. We contrast these results with those from our SHT simulations later.

We first note some additional details of the redox species used. In all studies we first made sure that essentially a single eigenstate of the redox species was involved in the ET, by choosing a state well isolated in energy. We then moved the energetic position of this redox species state (relative to the semiconductor conduction band edge) to an appropriate position for ET, by altering the redox species pseudopotential, as mentioned above. We denote the energy (eigenvalue) of this redox species state as ε_{ET}, and its magnitude shifts with the electostatic solvent potential acting on it. In our oxidation studies we prepared the electron density in this state initially, and in reduction the electron density coming from the semiconductor primarily tunneled into this state. This state, of course, is actually quasi-stationary when ε_{ET} is energetically positioned in a band and when the redox species is moved near the surface because the state is then electronically coupled to the continuum of semiconductor states. In this case the state is energetically broadened and has a finite lifetime, τ, via the time-energy uncertainty principle ($\tau \sim \hbar / |V|$, $|V|$ is the appropriate electronic coupling matrix element) (18-21). (Reference (2) discusses how our ε_{ET} relates to the Gerischer model "energy levels" (22).)

Let us first see what oxidation looks like pictorially in our simulations. In fig. 1

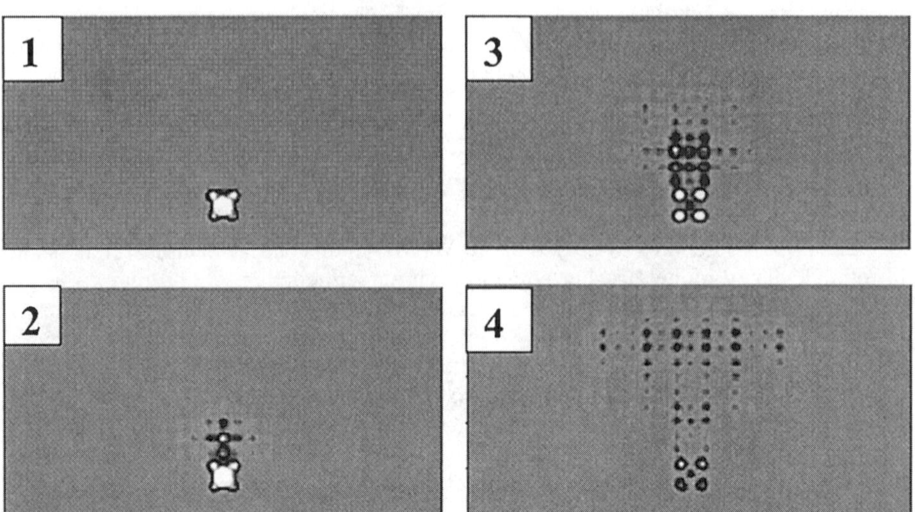

Figure 1. Isosurfaces of electron density in a cross section perpendicular to the surface and passing through the redox species. The numbered frames show successively longer times.

we see the electron probability density propagating from the redox species to the semiconductor bulk using the fixed trajectory method when ε_{ET} lay in the conduction band, the system was in a TS configuration, and the electron density was initially all in the redox species. Figure 1 shows the electron probability density isosurfaces in a cross section perpendicular to the surface. The density progressively leaks out from the redox species and into the semiconductor with time, as shown. We employed the second redox species described above (which traps significant density on each of its six oxygen atoms, with their modified pseudopotentials).

Frozen Solvent Oxidation

Let us denote by n_{REDOX} the electron probability density in the redox species. The behavior of this quantity when the frozen solvent method is employed is quite predictable. No density is exchanged with the semiconductor when ε_{ET} lies in the band gap, i.e., n_{REDOX} is 1 for all times. When ε_{ET} lies in the conduction band n_{REDOX} is zero in the long time limit, i.e., all the density escapes. Indeed, n_{REDOX} decays irreversibly and essentially exponentially in time in accord with well known analytical models for the decay of a discrete state into a continuum of states (*2, 18, 20*).

The above results reveal a well-known principle of quantum mechanics (*18, 23*) which governs the behavior of our system. In the SLI oxidation case, where an essentially discrete initial state is coupled to a continuum of final states, the system leaves the initial state *irreversibly* under the time independent potential conditions used. This behavior also markedly affects the ET in the time dependent potential case, as next described.

Fixed Trajectory Oxidation

In SLI ET the potentials and electronic states are, of course, time dependent via their coupling to the solvent, semiconductor, and redox species nuclear dynamics. Let us now look, in fig. 2, at results of our fixed trajectory oxidation calculations, which

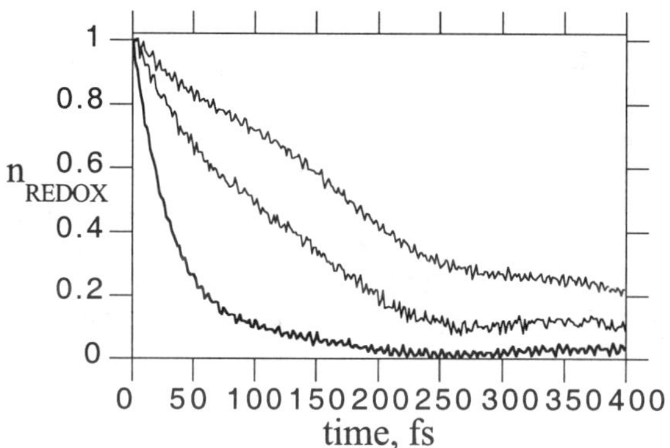

Figure 2. n_{REDOX} as a function of time for fixed trajectory oxidation simulations.

include this dynamics. With the potentials we chose, during a solvent fluctuation to the heterogeneous oxidation TS ε_{ET} moves from a point near the reduced species average equilibrium position (for simplicity we used redox species potentials which

72

placed this in the bandgap) to a position in the conduction band. The nuclear trajectory which produced the curves of fig. 2, for example, makes ε_{ET} rise from this equilibrium bandgap position into the conduction band then drop back into the band gap where ET is shut off and n_{REDOX} consequently levels off (see fig. 2). Fig. 2 shows that as the time spent by ε_{ET} in the conduction band (essentially the transition state lifetime, τ_{TS}) increases relative to the approximate lifetime of the redox species electronic state (τ_{EL}) the long time n_{REDOX} decreases (i.e., at 400 fs in fig. 2. the curves lie successively below each other at longer τ_{TS}). Complete transfer of the electron probability density can occur if $\tau_{TS} >> \tau_{EL}$. (This corresponds in traditional ET models to a transmission coefficient of one, and, essentially, the adiabatic limit (2).) τ_{EL} depends, of course, on the redox species distance from the surface and the position of ε_{ET} in the conduction band (roughly $\tau_{EL} \sim \hbar/|V|$)

In the case of a frozen solvent, we saw that redox species electron density leaked out exponentially in time. One might try to apply this to the dynamic solvent oxidation case by making the approximation that

$$n_{REDOX} \sim \exp(-t_c/\tau_{EL}) \tag{4}$$

where t_c is the time spent crossing the band of conduction band states in the trajectory. In the dynamic solvent case with the fixed trajectory method we sometimes find that roughly exponential decay of the state occurs in accord with eq. 4. However, the decay rate can change significantly with time as the coupling changes with position of ε_{ET} in the conduction band. Also, substantial quantum interference due to electronic coherence can arise. Thus we find in the dynamic solvent simulations that n_{REDOX} only sometimes follows the exponential decay of eq. 4; in fact the behavior can be very complex, in accord with our time dependent Anderson-Newns models of SLI ET (2).

SLI Reduction

We now examine reduction of redox species by electrons in the conduction band. That is we will examine propagation of a bulk electron through a crystalline lattice with ultimate scattering into the surface and transfer to a redox species. To better analyze the physics of this process we will examine the effect of propagating electrons of various energies toward the surface.

In this section we explore the reductive regime by using a Gaussian type wave packet for the bulk semiconductor electron formed by a superposition of Bloch states. If the wave packet utilized is larger than a certain size the results are approximately the same as for plane waves. (system specific size can be estimated from the Heisenberg uncertainty relation, $\Delta x \sim 1/\Delta k$, where Δk is representative of wavenumbers for conduction band semiconductor electrons.) We employ large wave packets below, but the wave packets are often small enough (of spatial scales of tens of angstroms) that some residual effects of size can remain. We are developing methods for examining larger spatial scales.

Let us again first present images of reduction in our simulations. In fig. 3 we show the results, via a cross sectional slice through the system, of propagating an electron wave packet from the semiconductor bulk to the surface in a fixed trajectory WPMD simulation using the second redox species described above (a typical thermal group velocity of 1.e7 cm/s was used for the wavepacket (24)) when ε_{ET} lay in the

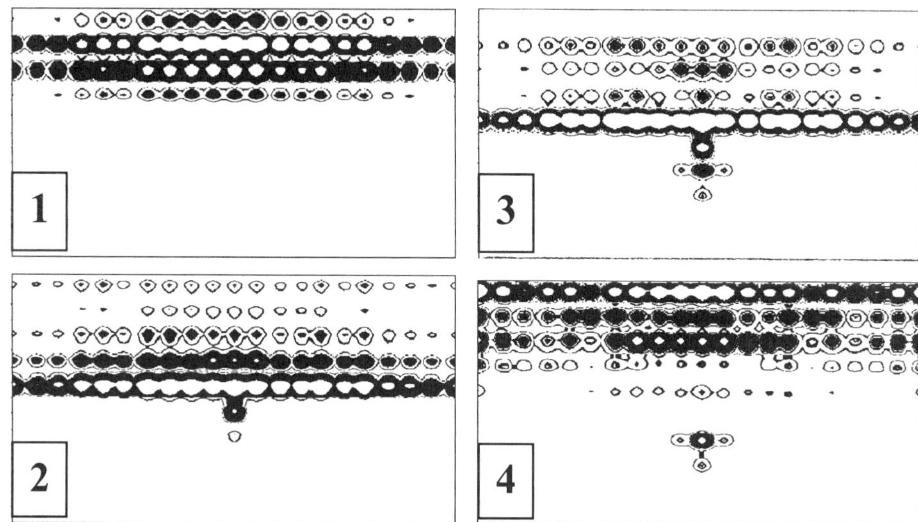

Figure 3. This figure is the same as fig. 1 except it shows reduction. The total elapsed time was about 300 fs. The wave packet initially approaches the semiconductor surface from the top of the frame; atoms of the redox species (bottom) become outlined by density.

conduction band. One can see part of the incident wave packet is transmitted to the redox species and part is reflected back to the bulk, which is of course familiar quantum behavior. A quasi-stationary state of the redox species (note some of its atoms become outlined with density) transiently traps some of the electron density which then leaks back into the semiconductor after a time delay, i.e., reemission occurs, which is also familiar behavior (*20, 21*). Some penetration of density into the solvent can be seen, which is short lived.

Frozen Solvent Reduction

Figure 4 plots n_{REDOX} as a function of time for various initial electron energies, in a frozen solvent simulation of reduction. This figure shows that the redox species is filled to various maximum occupation values in varying amounts of time as a function of the various semiconductor electron energies. The density is then reemitted back into the semiconductor in varying amounts of time. A rapid change in lifetime of the redox species state, as measured by its decay from the peak of the occupation profile, occurs with electron wave packet energy. This sensitivity to electron energy is, of course, the hallmark of resonant behavior (*18, 23*). On resonance both the wave function in the redox species, and the decay lifetime match that of a particular redox species eigenstate, i.e., the one we chose, see above. This proves the state occupied by the electron is a quasi-stationary state related to this eigenstate. Off resonance the occupation curves can of course show more complicated behavior than symmetric filling and decay (see fig.4).

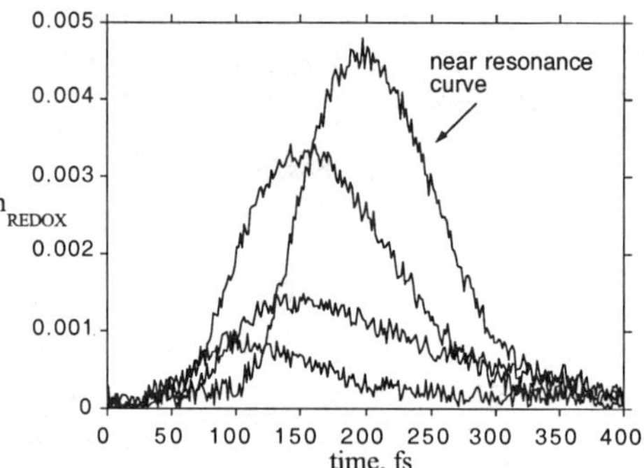

Figure 4. n_{REDOX} as a function of time from frozen solvent reduction simulations for various initial semiconductor electron energies (over a range around the resonance energy of about $\pm$ .025 eV).

Fixed Trajectory Reduction

The effect of solvent dynamics is very important in reduction, and we briefly examine it next with the fixed trajectory method.

In our fixed trajectory reduction simulations the nuclear dynamics forces ε_{ET} to change from the average equilibrium value of the oxidized ε_{ET} (situated in the conduction band) to a value in the bandgap (to the average reduced ε_{ET} value). We vary the electron wave packet impact time on the surface by starting the electron wave packet propagating at different times relative to the time ε_{ET} starts to change from its equilibrium value (propagation distance is fixed). Fig. 5 shows there is complicated interplay between the ε_{ET} dynamics and the electron dynamics. That is, to capture significant density we must have ε_{ET} near the resonant energy when the electron wave front is close enough to the surface to allow tunneling. Fig. 5 shows again (as in fig. 4) that the density in the redox species first increase in time, reaches a maximum, and then is "reemitted" back into the semiconductor, *but* the sinking of ε_{ET} into the bandgap can now stop the reemission process. If the redox species is close to the surface the emptying time back into the semiconductor will be short (via: $\tau_{EL} \sim \hbar/|V|$) and the solvent must bring the level quickly into the band gap to trap significant

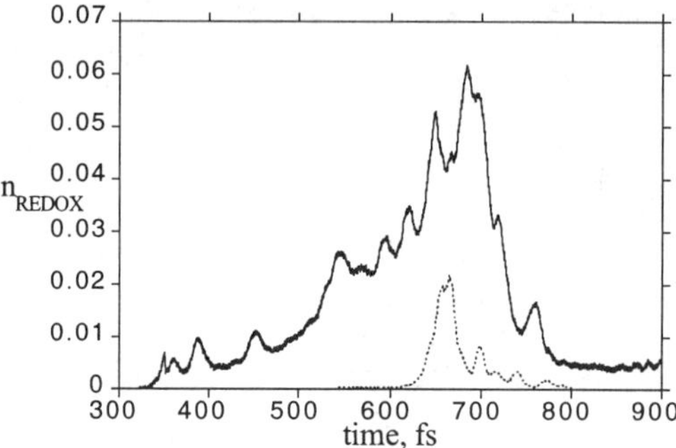

Figure 5. n_{REDOX} as a function of time for two fixed trajectory WPMD reduction simulations.

density. If it is far from the surface the emptying will be slow (as will the filling on resonance). Thus in this picture for a given rate of change of ε_{ET} caused by solvent dynamics there can be an optimal distance where the trapped density in the redox species is highest.

We are currently refining acetonitrile pseudopotentials and pseudopotentials for redox species used in that solvent. Our preliminary results are similar to those just discussed.

Discussion and SHT Comparison

The dynamics of ET at the semiconductor-liquid interface has been a subject of controversy for some time. The discussion below of the dynamics we see in our simulations significantly extends that we have given elsewhere (2), and clarifies aspects of this dynamics. We first discuss the dynamics in our frozen solvent and fixed trajectory simulations. We then compare this with the SHT WPMD methodology and results. We begin with the oxidation case.

For the SLI we expect irreversibility to characterize oxidation under most circumstances. For the frozen solvent regime in the oxidation case we saw irreversible exponential decay. This is well known behavior for discrete to continuum transitions, as explained above. We saw that the dynamic solvent fixed trajectory simulations can also exhibit exponential decay of the redox species density in time when ε_{ET} is in the conduction band. But more complicated behavior arises when the electronic coupling changes significantly with time and/or significant quantum interference arises. Another complication is that if energy is to be conserved in fixed trajectory simulations the solvent velocity or semiconductor modes (when we include lattice dynamics) must change and this affects the ET dynamics as well. But as long as the semiconductor is a perfect crystal (static and periodic) we expect irreversible behavior to characterize even the dynamic solvent case oxidation, i.e., no electron density should flow back into the redox species once it has left it and entered the semiconductor. Thus there will be reverse ET to the redox species from this outgoing wave packet entering the semiconductor essentially only by backscattering to the redox species via deviations from perfect crystal properties (defects, phonons, impurities etc.). We note that such imperfections, especially disorder at the solvent/solid interface, may typify real world systems. (The latter has interesting implications regarding Anderson localization phenomena). Regarding the phonon backscatter mechanism, a significant energy change would be involved so it would tend to be inefficient. (We held the semiconductor lattice frozen in the simulations of this paper, which eliminates phonon scattering.)

The results just discussed stand in stark contrast to the kind of SLI ET dynamics suggested by some conventional literature. For example, a number of papers adopt ideas from early heterogeneous ET models. These early heterogeneous models are closely related to traditional two-state state models of homogeneous ET between two molecules solvated in a liquid (22, 25, 26). In these homogeneous models, at the TS we have a resonance condition set up and if an electron wave packet is initially localized in an eigenstate of one molecule it will simply oscillate back and forth between the two molecules with a given frequency for as long as the TS lasts. These oscillatory reversible ideas have been presented as applying to the SLI.

If we look not at conduction band oxidation ET processes, but at ET to surface states in the band gap, reversibility like that in the two-state model might occur. The density of states for surface states in the bandgap can be narrow in energy width and thus approximately discrete. In this case reversible transfer might occur (we can see it in our WPMD), and a two-state model could be a first order description of it.

SLI conduction band reduction also does not in general exhibit reversibility in the sense used in two-state models. In the reduction case in the frozen solvent regime we saw (fig. 4 and 5) that when this occurs the redox species first fills with density and then it empties. This emptying, or back reaction, follows from the irreversible exponential decay we discussed. Thus, this back reaction (reverse transfer) has once again nothing to do with the kind of reversible behavior seen in the homogeneous two-state model, although this has also been suggested in the literature.

Further Details of the Methodology of the SHT Approach

When one examines the nuclear and electronic dynamics with SHT methodology the picture of SLI ET becomes many times more complicated. Our discussion of it, which we now present, will reveal further features of our lower level results in the last section. In traditional ET theory only two extreme regimes are described. One involves high electronic coupling, and the other very weak, and these correspond to the electronically adiabatic and, essentially, Fermi Golden rule limits, respectively (*25-29*). In the adiabatic case the Born-Oppenheimer (BO) approximation holds. This is the prevailing approximation used in ET simulations, and this consequently assumes that the nuclear forces are determined by the Born-Oppenheimer ground state electronic energy surface. However in general in all regimes other than this extreme one the nuclear forces are actually determined by numerous coupled electronic energy surfaces, and electronic transitions occur between these. The SHT simulations we use account for the latter nonadiabatic transitions.

The general procedure we use for our SHT simulations involves two methods: Tully's fewest-switches method (*5*) and Coker's hybrid of Tully's and the Webster et al. methods (*4*). In both methods an ensemble of trajectories is propagated over electronically adiabatic surfaces, while allowing for transitions between surfaces in the high electronic coupling regions. In these approaches if a transition to a new energy surface is made there is a nonzero energy gap which is traversed. The classical nuclear trajectory depends on the time evolution of the quantum electronic subsystem which itself depends on how the classical system changes in time. Thus the classical trajectory should be determined self-consistently. If this self-consistency is not fully addressed, as in Tully's approach, energy is not conserved and ad hoc methods must be employed to redistribute energy to the classical subsystem (random redistribution to solvent modes is the most primitive of these methods, as in our fixed trajectory method). Pechukas developed a rigorous semiclassical formalism for self-consistently propagating the classical and quantum subsystems, so that ad hoc adjustments are not required. Webster and co-workers developed a computationally feasible implementation for condensed matter systems of Pechukas' technique (*4, 6, 7*). But in Webster's approach the electronic dynamics is only propagated coherently only over a time step Δt. Coker's hybrid method, by combining Tully's and Webster's approach, allows for coherent propagation over all time and for self-consistent trajectory determination which eliminates ad hoc energy redistribution (*4, 6, 7*). The latter technique is the hybrid method we use, and we now explain additional details of the methods.

The SHT approaches we use contain a number of common features. As in all other methods reported in this paper, the nuclei are treated classically and only the electron involved in the ET is treated quantum mechanically. To perform a simulation we first assign the initial conditions (positions and velocities) for the nuclei. The initial state for the quantum subsystem (the ET electron) is prepared in whatever state is appropriate (localized on redox species or in the semiconductor). Coherent or incoherent mixed states can be prepared initially. We next obtain the instantaneous adiabatic eigenvalues and eigenvectors for each nuclear configuration at a time t by diagonalizing the electronic Hamiltonian.

There are many differing details of the steps which follow in the two SHT methods, and they have been extensively discussed (4-7, 30-33). We now briefly address some of these details. In Tully's method we next solve the equations of motion for the classical nuclei (4, 5, 30-33) for a time step Δt over the initially occupied potential surfaces for the assortment of ensemble members. Next the equation of motion (4, 5, 30-33), for the electronic subsystem is solved; this is an equation of motion for the quantum expansion coefficients of each ensemble member. (The expansion coefficients are the eigenvectors.) Then the hopping probabilities at the new nuclear configuration are computed (4, 5, 30-33). If no hop occurs the procedure repeats, if a hop does occur then the trajectory begins propagating over the new energy surface, and we redistribute energy by the method discussed above. In the hybrid approach Webster's linear approximation for the time dependence of the Hamiltonian during time Δt is used. Diagonalization of this Hamiltonian occurs at each small electronic time step n (at t+n δt) to obtain approximate eigenstates and eigenvalues in terms of the adiabatic basis which was obtained by the initial diagonalization (see above) at time t. These approximate eigenstates and eigenvalues are used to solve the equation of motion for the expansion coefficients for time Δt (4, 6, 7). The hopping probabilities are then computed via the same algorithm as in Tully's fewest switches method for each time t+n δt, and ultimately expansion coefficients are obtained to use in the Pechukas force (4, 9, 34) and the nuclei are then evolved under this force. The procedure is iterated to self-consistency. The propagation of the expansion coefficients is performed coherently from one time step Δt to the next, and the self-consistent iteration eliminates the need for ad hoc energy redistribution (4, 6, 7).

In both the SHT methods we employ, when the trajectory moves away from high coupling regions the dynamics is adiabatic with the appropriate percentage of ensemble members on each of the adiabatic surfaces, as determined by the quantum mechanical occupation probabilities. Thus, when a trajectory leaves a region of strong electronic coupling, where electronic transitions are highly probable, the nuclear dynamics becomes adiabatic and thus a function of a single Born-Oppenheimer energy surface. Such an approach correctly describes multichannel processes, where a number of outcomes are possible for a given scattering event. This is in contrast to some other SHT methods which propagate the dynamics on hybrid adiabatic surfaces, and which as a consequence contain physical anomalies.

Comparison of SHT and Fixed Trajectory Methodology and Results

The picture the SHT simulations paint is more complex that those we have so far discussed. When a region of strong electronic coupling arises the trajectory bifurcates and some probability density is ultimately associated with one adiabatic nuclear trajectory, and other density with other trajectories. All these trajectories obey different interatomic forces stemming in our simulations mainly from charge density changes of the redox species, which in turn affects the solvent dynamics. Thus we have to imagine a swarm of nuclear trajectories each of which contains some fraction of the electron probability density and the fraction in each and its structure changes with time. This is the kind of mental picture that emerges from all models of the SHT type.

Let us now compare the conceptual framework and methodology of the fixed trajectory and SHT simulations, using the oxidation case to illustrate some basic points. Before the excursion to the TS the redox species is in its reduced state. After the TS excursion the redox species either remains in its reduced state or it is oxidized. Thus, in the real world, from the perspective of the solvent (and redox species) nuclei, after the excursion they see either a reduced or oxidized species and their dynamics is affected accordingly. The trajectory for ε_{ET} we considered above for the fixed trajectory oxidation case, that is, (bandgap$\rightarrow$conduction band$\rightarrow$bandgap) represents an unsuccessful nuclear trajectory for oxidation, i.e., the nuclear trajectory relaxes to its initial conditions (equilibrium with the reduced redox species). A successful trajectory would ultimately end up in the conduction band (our equilibrium position of ε_{ET} for the oxidized species), relaxing to equilibrium with the oxidized species. In the fixed trajectory WPMD method, in effect, only crude approximate guesses of the actual trajectories the system follows in its ascent and descent from the TS are used (vide supra). Only in the Webster-Tully hybrid method is it possible to actually calculate these trajectories by including the main aspects of the self-consistent nuclear and electronic system coupling that determines them. To do so we place the chosen initial electron wavepacket in a TS configuration just as in the fixed trajectory method. Iterations of the Webster algorithm are then started which finally find the self-consistent trajectories both forward and backward in time relative to the initial TS configuration. This final trajectory evolves self-consistently with a proper accounting of the dynamical mixture of states, resulting in an appropriate statistical distribution of occupied states.

Thus with Coker's Tully-Webster hybrid method the average lifetime of the transition state for the various nuclear trajectories is calculable, whereas it is, in effect, only a highly approximate adiabatic TS lifetime that emerges from the fixed trajectory method (vide supra), and the statistical distribution of trajectories is different. The current mundane limitation to the Tully-Webster hybrid technique, which we are working on to improve, is that it is only possible to track a small number of trajectories due to computational demands. The trajectories we have been able to generate do seem to typically agree qualitatively with those of the Tully method in the ET regimes that we have so far examined. But differences in the trajectories of the two methods are inevitable due to the ad hoc energy redistribution of the Tully method.

Despite the shortcomings of the fixed trajectory calculations, we find, so far, that they do show behavior qualitatively similar to the SHT simulations. For example, consider the fixed trajectory oxidation simulations. When the redox species state ε_{ET}

reaches one of the continuum of conduction band states part of the electron probability density tunnels into that state in the semiconductor and part remains on the redox species. Some density also is transferred to other states in the continuum in the energetic vicinity of that state. The wavepacket in the semiconductor indeed becomes a superposition of these states, similar to the superposition in the SHT approach. Obviously the composition of the occupied states and the phase coherence in the two methods will be different. If we use an unsuccessful nuclear trajectory for a fixed trajectory oxidation simulation (bandgap→conduction band→bandgap) some density will typically remain in the redox species after the TS excursion. The amount of density left in the redox species, in effect, predicts the percent of systems having identical nuclear trajectories which will produce a reduced species after the TS excursion. The percent which will produce an oxidized species after the excursion is predicted by the amount of probability density which enters the semiconductor. These predictions can be compared with SHT results, and we have found qualitative agreement. Obviously the ensemble averaged transmission coefficients so obtained can also be compared to traditional ET theory. Also we see in both the SHT and fixed trajectory techniques the ubiquitous SLI irreversibility, that the amount of density transferred depends on the strength of the electronic coupling and the time spent in the high coupling regions, that in reduction the state first fills with and then empties its density. In short, while the methodology and results of our SHT simulations are much more complex than the fixed trajectory approach, at this stage in our studies it seems the same qualitative behavior often emerges.

Since the same basic qualitative trends are evident in both simulation methods (fixed trajectory and SHT) it is apparent that either method can be used to examine the basic physics of this system. This kind of similarity is seen for a wide variety of physical systems examined with these two basic methods. Sometimes this similarity is nearly quantitative, as with the simple two-state Landau-Zener model (which assumes a fixed trajectory) and higher level simulations for two-state systems *(10, 35-37)*. Our system is so complex that quantitative agreement is not at all to be expected. Indeed, as we saw, the fixed trajectory method cannot address some of the most important quantities calculated by the Coker-Webster-Tully hybrid method (e.g., the self-consistent trajectories themselves). Yet the simplicity of the fixed trajectory method can greatly aid the physical understanding of the SLI, as illustrated above. We have shown other examples of the latter elsewhere *(2)*.

Comparison to Traditional Outlooks

A natural question to ask is where traditional theory stands as contrasted to the outlooks we have presented. For example does the often employed Landau-Zener (LZ) model of ET apply to the SLI system? The LZ model assumes a fixed nuclear trajectory *(19, 36-39)*. Thus, as we have seen, one cannot expect it to model nonadiabatic effects in any detail, even though it has been employed over the entire range of electronic coupling from very weak to very strong. The next question that arises is whether it can model SLI ET transitions, i.e., transitions from a discrete state to a continuum or vice versa. We have shown elsewhere *(2)* that there are grave difficulties with this model regarding the latter. The main difficulty is that LZ theory is a two-state model; but, as we have seen, clearly the SLI involves a continuum of states, and discrete/continuum transitions. When one attempts straightforward generalizations of this two-state model to the continuum case, as we did in ref. *(2)*, a number of serious anomalies arise (*2*). Finally LZ theory does not account at all for quantum interference, which can be very important in the SLI case.

Final Remarks

Traditional ET theory can only address two extreme regimes of ET (electronically adiabatic and the weak coupling Fermi Golden Rule limit). Further, the detailed complexities of discrete-continuum transitions have not been addressed in it. In light of this it seems important to explore new avenues in theory building to model both real world SLI systems and new computer simulations. It appears WPMD in its SHT form should offer a rewarding approach to address both the nonadiabatic ET regimes intermediate to those just mentioned, and the details of the complicated electronic discrete-continuum transitions of the SLI system. The WPMD approach, as well as others we have explored (*13, 14, 40, 41*), seems to us a better candidate for addressing and conceptualizing the multi-state heterogeneous problem than many alternative approaches. We believe our results above show the WPMD model offers a detailed picture of the SLI ET mechanism, and that it is capable of addressing much of the SLI's rich dynamics. Further, it is applicable to most of the new systems of interest in the SLI community, can explore the limitations of small scale simulations, and can be couched in a scattering theory cross section perspective.

While our WPMD model clearly has advantages, a one-electron model has obvious shortcomings in the ultimate quest for a complete SLI model. High level many electron correlation and exchange methods must be used for energetically accurate and complete descriptions. These methods can be employed in first principles MD simulations, like those we are pursuing (*13, 14*). The challenge for the future lies in making high level SLI ET calculations computationally tractable, and also in developing conceptually helpful models for this complex system. It is our belief that the WPMD model takes steps toward meeting these two challenges.

Acknowledgement. This work was supported by the U.S. Department of Energy, Office of Basic Energy Sciences, Office of Energy Research, Division of Chemical Sciences.

Literature Cited

1. Miller, R. J. D.; McLendon, G. L.; Nozik, A. J.; Schmickler, W.; Willig, F. *Surface Electron Transfer Processes*; VCH Publishers: New York, 1995.
2. Smith, B. B.; Nozik, A. J. *J. Phys. Chem. B.* **1999**, *103*, 9915.
3. Cohen, M. L.; Chelikowsky, J. R. *Electronic Structure and Optical Porperties of Semiconductors*; Springer-Verlag: New York, 1988; Vol. 75.
4. Coker, D. F. *Computer Simulation Methods for Nonadiabatic Dynamics in Condensed Systems*; Allen, M. P. and Tildesley, D. J., Ed.; Kluwer, 1993, pp 315.
5. Tully, J. C. *J. Chem. Phys.* **1990**, *93*, 1061.
6. Webster, F.; Rossky, P. J.; Friesner, R. A. *Comput. Phys. Commun.* **1991**, *63*, 494.
7. Webster, F. J.; Schnitker, J.; Friedrichs, M. S.; Friesner, R. A.; Rossky, P. J. *Phys. Rev. Lett.* **1991**, *66*, 3172.
8. Baer, M. ; CRC: Boca Raton, 1985; Vol. II, pp 219.
9. Pechukas, P. *Phys. Rev.* **1969**, *181*, 166.

10. Nakamura, H. *J. Chem. Phys.* **1987**, *87*, 4031.

11. Dehareng, D. *Chem. Phys.* **1986**, *110*, 375.

12. Sawada, S.; Metiu, H. *J. Chem. Phys.* **1986**, *84*, 227.

13. Smith, B. B.; Nozik, A. J. *J. Phys. Chem. B* **1997**, *101*, 2459.

14. Smith, B. B.; Nozik, A. J. *Chem. Phys.* **1996**, *205*, 47.

15. Curtiss, L. A.; Halley, J. W.; Hautman, J.; Nagy, Z.; Rhee, Y.-J.; Yonco, R. M. *J. Electrochem. Soc.* **1991**, *138*, 2032.

16. Kuharski, R.; Bader, J.; Chandler, D.; Sprik, M.; Klein, M.; Impey, R. *J. Chem. Phys.* **1988**, *89*, 3248.

17. Hautman, J.; Halley, J. W.; Rhee, Y.-J. *J. Chem. Phys.* **1989**, *91*, 467.

18. Cohen-Tannoudji, C.; Diu, B.; Laloe, F. *Quantum Mechanics*; John Wiley and Sons: Paris, 1977.

19. Landau, L. D.; Lifshitz, E. M. *Quantum Mechanics*; Pergamon Press: New York, 1977.

20. Schiff, L. I. *Quantum Mechanics*; 3rd ed.; McGraw-Hill: New York, 1968.

21. Taylor, J. R. *Scattering Theory: The Quantum Theory of Nonrelativistic Collisions*; Robert E. Krieger Pub. Co., Inc.: Malabar, Florida, 1987.

22. Morrison, S. R. *Electrochemistry of Semiconductors and Oxidized Metal Electrodes*; Plenum Press: New York, 1980.

23. Bohm, D. *Quantum Theory*; Dover: New York, 1951.

24. Sze, S. M. *Physics of Semiconductor Devices*; John Wiley and Sons: N.Y., 1981.

25. Marcus, R. A. *J. Chem. Phys* **1964**; Vol. 15, pp 155.

26. Marcus, R. A. *J. Chem. Phys.* **1965**, 679.

27. Levich, V. G. ; Delahay, P., Ed.; Interscience: New York, 1966; Vol. 4, pp 249.

28. Levich, V. G. *Physical Chemistry: An Advanced Treatise in Physical Chemistry*; Academic Press: New York; Vol. 9B.

29. Dogonadze, R. R. *Reactions of Molecules at Electrodes*; Wiley: New York, 1971.

30. Tully, J. C.; Preston, R. K. *J. Chem. Phys.* **1971**, *55*, 562.

31) Tully, J. C. ; Nonadiabatic Processes in Molecular Collisions in Dynamics of Molecular Collisions, ed. W.H. Miller, Plenum: New York, 1976, pp 217.

32. Tully, J. C. *Phys. Rev. B* **1977**, *16*, 4324.

33. Tully, J. C. *J. Chem. Phys.* **1980**, *73*, 1975.

34. Pechukas, P. *Phys. Rev.* **1969**, *181*, 174.

35. Chapman, S. *Adv. Chem. Phys.* **1992**, *82*, 423.

36. Gislason, E. A.; Parlant, G.; Sizun, M. *Adv. Chem. Phys.* **1992**, *82*, 321.

37. Nakamura, H. *Adv. Chem. Phys.* **1992**, *82*, 243.

38. Landau, L. D. *Physik. Zeitschr. Sowjetunion* **1932**, *2*, 46.

39. Zener, C. *Proc. Royal Soc.* **1932**, *A137*, 696.

40. Smith, B. B.; Hynes, J. T. *J. Chem. Phys.* **1993**, *99*, 6517.

41. Smith, B. B.; Halley, J. W. *J. Chem. Phys.* **1994**, *101*, 10915.

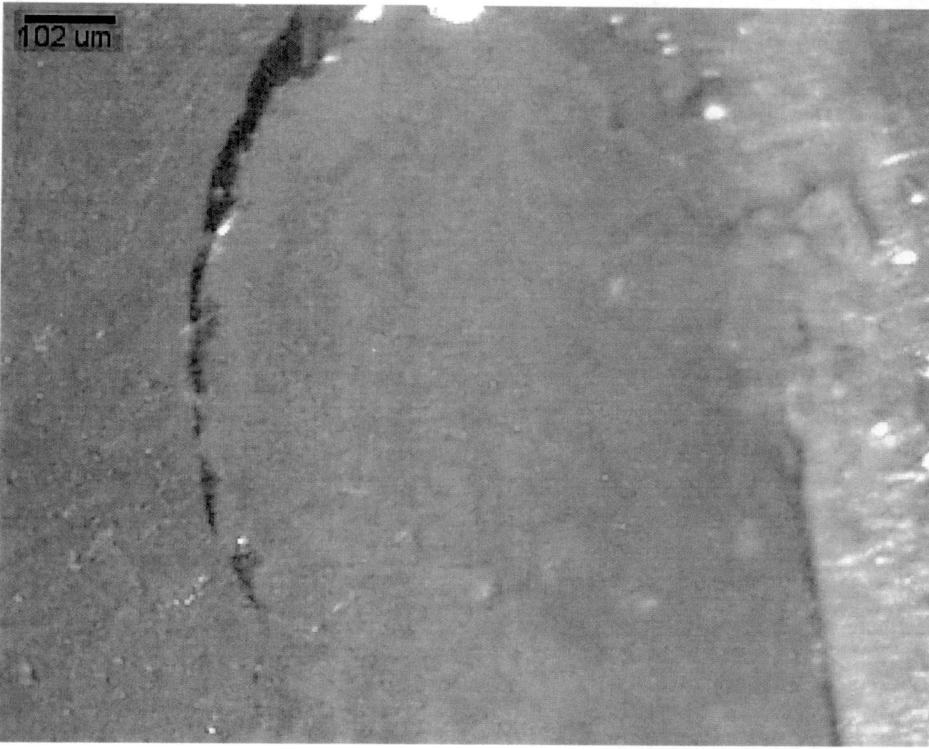

102 um

Oxides at Liquid–Solid Interfaces

Chapter 7

An X-ray Diffraction Study of the Passive Oxide Film on Iron

Mary P. Ryan[1], Michael F. Toney[2], Lucy J. Oblonsky[3], and Alison J. Davenport[4]

[1]Department of Materials, Imperial College of Science, Technology and Medicine, Prince Consort Road, London SW7 2BP, United Kingdom
[2]Almaden Research Center, IBM Research Division, 650 Harry Road, San Jose, CA 95120
[3]DuPont Central Research and Development, P.O. Box 80323, Wilmington, DE 19880
[4]School of Metallurgy and Materials, The University of Birmingham, Edgbaston, Birmingham B15 2TT, United Kingdom

The atomic structure of the passive oxide film grown on an iron electrode in near-neutral solution has been explicitly determined. The film is a highly defective, nanocrystalline spinel that is not identical to any bulk phase. This new phase has a fully occupied oxygen lattice with an octahedral site occupancy of $80 \pm 10\%$, tetrahedral site occupancy of $66 \pm 10\%$, and octahedral interstitial site occupancy of $12 \pm 4\%$. The passive film forms with an epitaxial relationship to the substrate iron. The in-plane lattice parameter for this phase (the LAMM phase) is 8.39 ± 0.01 Å and the out-of-plane lattice parameter is 8.3 ± 0.1 Å. The structure of the film formed at +0.4 V MSE and evaluated in situ was compared to the structure of the film formed at +0.4 V MSE and evaluated ex situ. The passive film that forms at this potential is stable to emersion, demonstrated by the identical structures of the oxides in the two cases.

Most metals are thermodynamically unstable to oxidation in a moist environment and can only be used industrially because the rate of this reaction is kinetically hindered by the presence of surface oxide films (passive films). The stability of engineering alloys is thus dominated by the characteristics of the oxide film with respect to dissolution, chemical or mechanical breakdown, and the ability of such oxides to repassivate the surface after disruption. Due to the technological importance of these films there has been considerable effort expended to unravel the details of their structure and chemistry. The formation of passive films is an electrochemical process that takes place in an aqueous environment and so in order to study the environmental effects on the film structure it is necessary to perform the measurements in-situ. This, coupled to the fact that these films are very thin, typically only a few nanometres thick, has made their study problematic. Along with this difficulty goes a long standing controversy as to the validity of conducting ex-situ studies i.e. is the film changed on removal from the aqueous conditions in which it was formed? The question of passive film structure has in particular been contentious and until recently there was argument about issues as basic as whether the film is amorphous or crystalline.

The study of the passivity of iron has a long history dating back to the work of a number of 18th and 19th century authors including Keir, Faraday and Schoenbein [1-3] who carried out elegant experiments to demonstrate the electrochemical nature of the process, and correctly attributed it to the formation of a "superficial film of oxide".

Using a method developed by Evans to isolate the oxide film from its metal substrate [4] Iitaka obtained a diffraction pattern from the passive film [5], showing it to be composed of either γ-Fe_2O_3 or Fe_3O_4. From chemical tests (details were not given), they concluded that the film was γ-Fe_2O_3, "not α-Fe_2O_3, contradicting the views of many previous workers".

Electron diffraction and electrochemical experiments in the 1960's [6,7] led to the suggestion that the film is crystalline and consists of an inner layer of Fe_3O_4 and an outer layer of γ-Fe_2O_3. This bilayer model has prevailed until recently. However, since the electron diffraction measurements were conducted ex-situ and in vacuum, the film structure may have been altered through crystallisation and/or removal of hydroxide groups (OH^-). These concerns led to studies of the passive film in the 1980's using a variety of in-situ techniques, for example, Mössbauer spectroscopy[8,9], extended X-ray absorption fine structure (EXAFS)[10,11], and surface enhanced Raman (SERS)[12,13]. These were interpreted, in contrast, as showing that the passive film is amorphous and/or similar to Fe hydroxides or oxyhydroxides. In contrast to this, recent XANES data[14] suggest that the passive film is either a spinel structure (such as Fe_3O_4 or γ-Fe_2O_3), or a highly disordered (amorphous) film.

The crystalline nature of the passive film on iron was confirmed in situ by Ryan et al using scanning tunnelling microscopy (STM)[15]. The data, which only measure the surface structure, were consistent with a spinel structure, but could not distinguish between Fe_3O_4, γ-Fe_2O_3 or other spinel-like variants.

The specific role of the oxide structure in passivity has been widely discussed in the literature. The advantageous effect of adding chromium to iron to form stainless steel has been attributed to the formation of a 'glassy' oxide film with 'self-healing properties'[11,16]. More importantly perhaps is the specific role that defects in the oxide film may play in film growth and film breakdown. Kirchheim[17] has discussed a model for film growth based on defect mediated transport processes that correlates the measured electrochemical currents to the assumed defect concentration. The point defect model, developed by MacDonald et al [18], describes both passivation transients and the critical parameters in oxide film breakdown. Given the controversy surrounding the oxide film structure it is hardly surprising that observation of defects in films have not been elucidated and that these models have not been rigorously tested. In order to understand and model these dynamic effects it is necessary to know both the film's crystal structure and to have a detailed picture of the defects present in these films.

To address these issues we have used synchrotron based x-ray diffraction to study, in-situ, passive films formed on pure iron with the intention of explicitly describing both the atomic structure and the nature of any defects present.

Experimental Procedure

Sample Preparation

Iron single crystals, 10 mm in diameter and 2-3 mm thick, with approximate orientation of (100) and (110) were obtained from Monocrystals Incorporated. They were oriented to within 1° and metallographically ground using SiC paper and polished with progressively finer diamond abrasive to reach a final surface finish of 0.25 μm. The crystals were then sputter-annealed in an argon ion beam for at least 200 hours at a temperature of 500°C and cooled to room temperature under vacuum. In addition, crystals were sputter-annealed immediately prior to use in experiments for at least eight hours.

Electrochemical Control

The experimental approach for in situ diffraction measurements was based on a procedure developed at IBM[19]. The use of a well-designed electrochemical cell allows accurate control of electrode potential whilst enabling surface studies to be performed simultaneously. A schematic of the cell used cell is shown in Figure 1. The single crystal sits at the centre of the cell, held by a magnet that also provides electrical connection. Ports in the cell allow insertion of a platinum counter electrode and an oxidised silver wire reference electrode (Ag/AgO). The reference was previously calibrated against a standard mercury sulphate reference (MSE) against which all potentials are quoted. The Prolene cell window can either be inflated by the addition of more solution through valves or deflated (sucked down) as shown in Figure 1.

When high electrochemical currents were required, for example during cathodic reduction or during the initial few seconds of passive film growth, the window was inflated to facilitate transport to the electrode surface. For diffraction measurements the window was deflated on to the surface of the electrode by removal of solution from the cell. This procedure left a layer of solution adjacent to the surface that was estimated to be a few μm in thickness. This procedure permitted maintenance of electrochemical control but minimised diffuse scattering of the X-ray beam by solution. An outer chamber (not shown in the Figure) constructed with Kapton® windows was filled with flowing nitrogen to prevent possible oxygen diffusion through the Prolene® window.

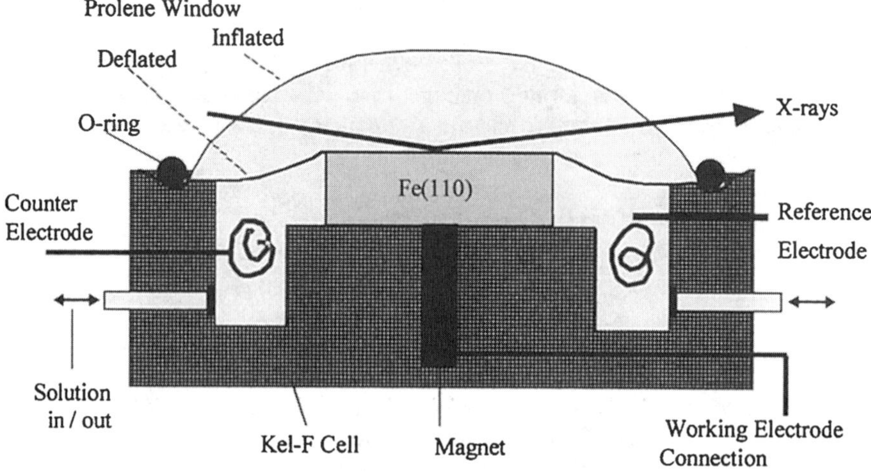

Figure 1 Electrochemical cell used for in-situ diffraction measurements

The solution used was a pH 8.4, 0.136 M borate buffer (7.07 gl^{-1} H$_3$BO$_3$ + 8.17 gl^{-1} Na$_2$B$_4$O$_7$.10H$_2$0) made from analytical grade reagents and 18 MΩcm deionized water. For each experiment, the sample was initially polarised cathodically, with the window inflated, to remove the native air formed oxide film. The potential was initially held at -1.8 V for 10-60 minutes and then set at -1.4 V MSE. This potential is low enough to prevent oxidation of the surface but with minimum production of hydrogen gas. Any hydrogen bubbles that had formed were flushed from the cell. To confirm that any air-formed oxide was fully removed during cathodic reduction, diffraction measurements were carried out on the reduced surface at -1.4 V with the cell deflated. The window was then re-inflated and the sample passivated at +0.4 V for one hour. This potential is 'high' in the passive region where it has been shown that passivation occurs without any prior dissolution of the metal surface[20]. Thus the measurements could not be complicated by the precipitation of iron containing species form the solution.

For ex-situ measurements the crystal was passivated for one hour in a hanging-meniscus geometry using a conventional three-electrode cell following the same cathodic treatment described above. The crystal was then emersed, rinsed in deionized water, dried in a stream of nitrogen and mounted on the spectrometer. All ex-situ measurements were made in an atmosphere of flowing high purity helium.

Diffraction Measurements

X-ray scattering data were collected at the National Synchrotron Light Source, Brookhaven National Laboratory, beam line X20A, with a focused X-ray beam. Several experimental runs were used to collect the complete data set from the Fe(110) and Fe(001) samples, and in all these runs an incident X-ray energy of about 10 keV (wavelength of about 1.24 Å) was used. When possible, grazing incidence geometry was used with an incidence angle of about 0.55° in order to minimise diffraction from the substrate. Otherwise, a symmetric geometry was used. The diffracted beam was analysed with 2 milliradian (mrad) Soller slits and the acceptance out of the scattering plane was about 15 mrad. A Si(Li) or Ge detector (depending upon availability) was used to discriminate the elastic scattering from the Fe fluorescence. For each diffraction peak, three measurements were made: an in-plane radial scan, a rocking or phi scan, and an 'L scan'. Data from these scans were then used to calculate the integrated intensities and obtain the structure factors used for the crystallographic analysis.

Results

The passive film was examined in situ during growth, and ex-situ to determine its structural stability on removal from the electrolyte. Initial in situ and ex situ measurements established that the passive film formed at +0.4 V MSE was in both cases a crystalline spinel as reported previously [21]. This was indicated by the symmetry of the oxide, its lattice parameter, the allowed reflections and, qualitatively, the diffraction intensities (Figure 2). Typical diffraction data obtained for the passive oxide are shown in Figure 3. The measured diffraction data are inconsistent with the spinels magnetite and maghemite (Fe_3O_4 and γ-Fe_2O_3) as demonstrated in Figure 3 which shows that these phases are not present in the passive film on iron. To more quantitatively determine the passive film structure we used the crystallographic structure factors for the 68 measured, symmetry inequivalent diffraction peaks (ref 21). With these the goodness of fit parameter for Fe_3O_4 and γ-Fe_2O_3 are unacceptably large (4.2 for Fe_3O_4 and 7.3 for γ-Fe_2O_3), further demonstrating that these phases are not present in the passive film. We note that these data also show that the film structure was identical (to within experimental error) in-situ and ex-situ i.e. the film formed at this potential is stable to emersion and exposure to laboratory atmosphere.

The lattice parameters of the passive film structure are 8.39 ± 0.01 Å in-plane and 8.3 ± 0.1 Å out-of-plane. This compares to 8.394 Å for Fe_3O_4 8.3396 Å and 8.3221 Å[22], for γ-Fe_2O_3. To determine the correct passive film structure, we first note that the lattice constants of the film are close to those of bulk Fe_2O_3 and Fe_3O_4. Since the lattice constants of spinel oxides are predominantly determined by the packing of oxygen anions[22], this suggests that all the oxygen sites are fully occupied. However, cation vacancies and interstitials in spinels have a low free energy of formation[23],

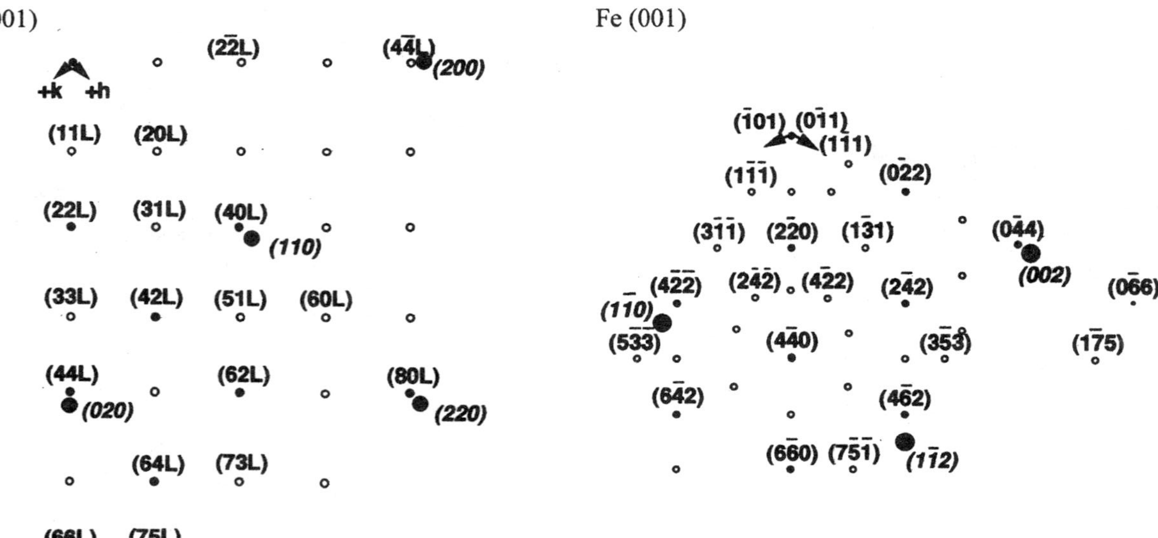

Figure 2. One Quadrant of the Diffraction Pattern for the Passive Film on Fe(001) and Fe(110) For clarity only some of the out of plane peaks are shown and are collectively denoted by 'L'. The diffraction peaks from the oxide are denoted by small filled and open circles (which are in-plane peaks). Substrate peaks are denoted by the large shaded circles.

and thus, we considered models with random cation vacancies and interstitials. Such a model (as described below) is found to fit the experimental data very well giving a χ^2 of 1.3.

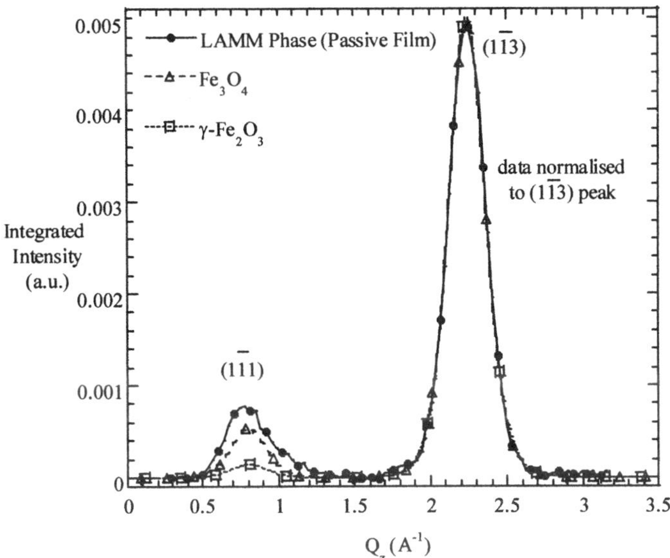

Figure 3. A radial scan comparing diffraction data from the passive film with that calculated form model Fe_3O_4 and γ-Fe_2O_3. Notice that the ration of intensities observed for the passive film can not be achieved by any combination of the two common spinel iron oxides.

The Atomic Structure and Point Defects

The spinel unit cell contains 32 oxide anions, and 16 octahedral and 8 tetrahedral cation sites [22]. In Fe_3O_4, these sites are fully occupied, while in Fe_2O_3, 25% of the octahedral sites have a 33% occupancy, and the other octahedral sites and all the tetrahedral sites are fully occupied. These spinels contain no interstitials. The structure of the passive film (Figure 4) has an octahedral site occupancy of 80±10% and a tetrahedral site occupancy of 66±10%; there are cations occupying 12±4% of the available octahedral interstitial sites, but no tetrahedral interstitials (Table 1).

The distribution of defects throughout the films has not been explicitly studied in the present work although future work is planned to probe the profile the defect structure. There are two main factors that are likely to produce a geometrically constrained defect structure. The first is defect clustering; it is likely that there is a correlation between tetrahedral vacancies and octahedral interstitials via the formation of energetically stabilised clusters. The second factor is the presence of a strong

92

electric field across the oxide. This field can be on the order of 10^6 V cm^{-1}. It is likely that the iron cations will be in the higher oxidation state (3+) close to the electrolyte i.e. vacancies are likely near the oxide-solution interface while interstitials are likely near the metal oxide interface. At present it is not possible to say which of these factors may dominate. For simplicity, our model (the LAMM phase) assumes a uniform distribution of vacancies and interstitials.

The stoichiometry for the best-fit structure is found to be $Fe_{(1.9\pm0.2)}O_{(3)}$, This means that most of the Fe cations are in the Fe(3+) oxidation state, consistent with previous in-situ experiments on passive film chemistry [20].

Table 1 Summary of passive film structure versus the spinel oxides

	octahedral cation site occupancy	tetrahedral cation site occupancy	anion site occupancy	Number of tetrahedral interstitials	Number of octahedral interstitials
Fe_3O_4	100	100	100	0	0
γ-Fe_2O_3	83 (ordered)	100	100	0	0
Passive Film LAMM phase	80 (random)	66	100	0	12

Fig. 4. Schematic illustration of the passive film (LAMM) structure. For clarity, only the bottom half of the unit cell is shown. The oxygen anions are shown by the large spheres (fully occupied), the cation sites by medium-sized spheres (tetrahedral 66% occupancy and octahedral sites 80% occupancy). Four of the 8 octahedral interstitial sites are shown by the small spheres in the bottom left of the figure; there are four equivalent sites at the upper right hand section, which are not shown.

Grain Size and Extended Defects

As well as the details of the atomic structure evidence of finite grain size, stacking faults and antiphase boundaries can be found in the diffraction peak widths. A summary of extended defects is shown in Table 2. The in plane particle size is 80 and 60 Å for oxides on Fe(001) and Fe(110) respectively. The films are thus categorised as *nanocrystalline.* The corresponding out-of-plane grain size (surface normal) was 35 and 25 Å, which is on the order of the film thickness. Such a small crystallite size may help to explain why the spinel rather then the thermodynamically stable a-Fe_2O_3 is formed. It has been shown for alumina that surface energy considerations can stabilise γ-Al_2O_3 over the thermodynamically stable α-Al_2O_3 [24]. The effect of this fine grain size on the film's properties as yet unknown although it is likely to be significant. Nanocrystalline materials are increasingly recognised as having mechanical properties (e.g., hardness and ductility) that are distinctly better than bulk solids[25]. By conferring mechanical robustness, the film's nanocrystalline nature may be important for its corrosion resistance. Also, the transport mechanisms in nanocrystalline materials are likely to be strongly affected by the large number of atoms residing along grain boundaries. It is estimated that between 15 and 50% of atoms in a nanocrystal are within one atom distance of the grain boundary and are thus significantly relaxed from the normal lattice position.

Evidence for stacking faults along the (111) direction was observed for the Fe(110) crystal. Similarly, evidence of antiphase boundaries was obtained for passive films formed on both the Fe(001) and Fe(110) crystals by plotting peak width versus diffraction order for the [h00] and [hh0] peaks in Fe(001) and [h$\bar{h}$0] in Fe(110) as shown in Figure 5

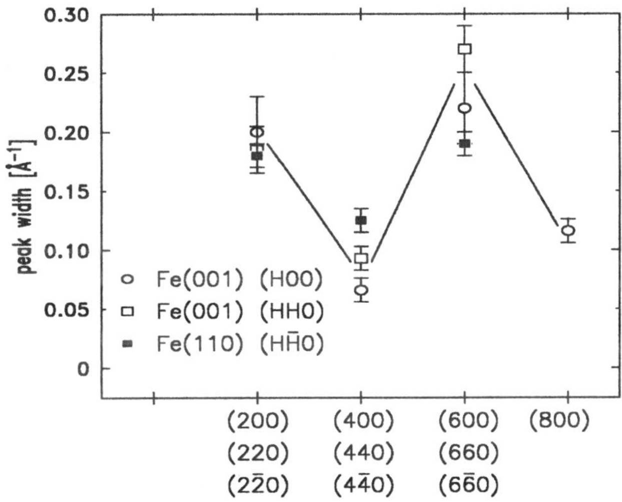

Figure 5. Comparison of peak widths for a series of in-plane scans. The 'zigzag' pattern is indicative of the presence of antiphase boundaries.

Table 2. Summary of Extended Defects in the Passive Film Structure

Defect		Fe(001)	Fe(110)
Particle Size 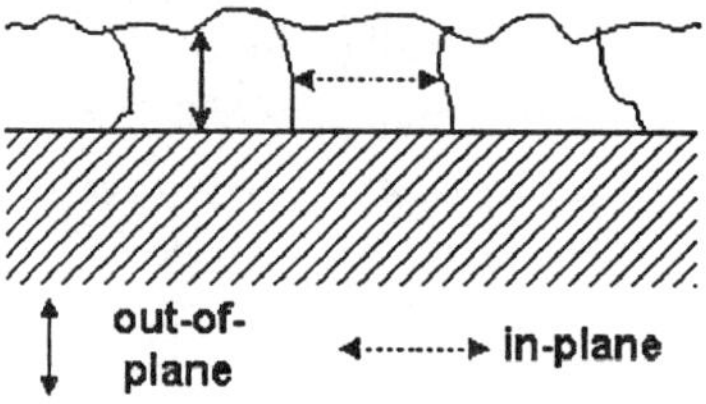 — out-of-plane, in-plane		~35 Å, ~80 Å	~25 Å, ~50 Å
Mosaic Spread 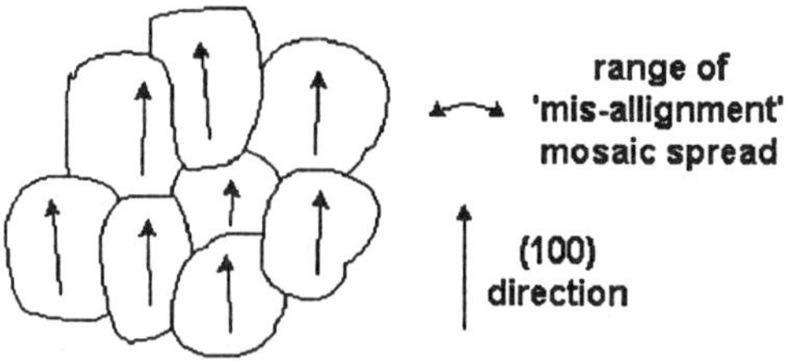 — range of 'mis-allignment' mosaic spread, (100) direction		2.5 °	4.1 °
Stacking Faults 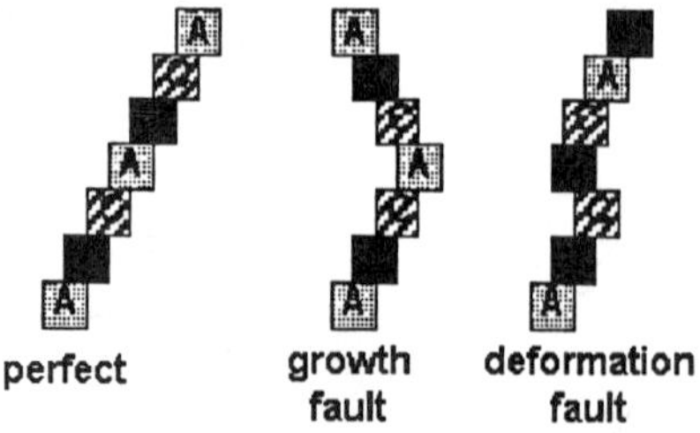 — perfect, growth fault, deformation fault		???	yes along (111)
Antiphase Boundries 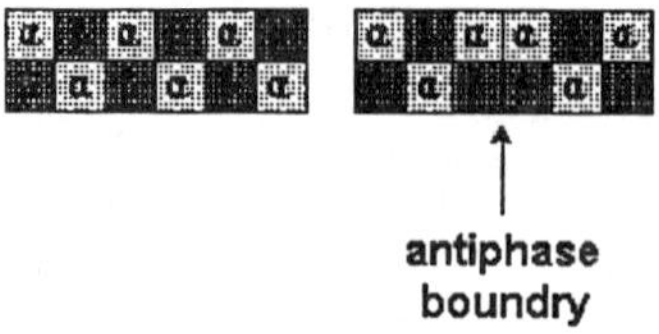 — antiphase boundry		~ 20 Å	~ 25 Å

Epitaxy and Mosaic Spread

The passive film grows with a well-defined orientation relationship to the substrate such that the (001) oxide plane grows parallel to an Fe(001) surface and the oxide [1 $\bar{1}$ 0] direction parallel to Fe[100]. The orientation relationships found between the oxide and the metal substrate appears to be dominated by the similarity between the oxide (220) and Fe (100) interplanar spacings (2.89 and 2.96 Å respectively). This leads to the observation on diffraction patterns of coincidence between Fe(200) and oxide (440) reflections, and Fe(110) and oxide (400) reflections. It also leads to zone axes (directions) of the form Fe<100> and Fe<110> to be parallel to oxide<110>, and oxide<100> to be parallel to Fe<110>. The observation that the (111) oxide planes grow parallel to a Fe(110) surface is also consistent with previous STM observations of the passive film[15], as well as electron diffraction of passive films[26] and for the air-formed film[27]. Despite this strong orientation relationship broadening of the phi-scan peak due to mosaic spread was observed. This is a measure of the degree of 'misalignment' of the individual grains. From the peak widths the mosaic spread is 2.5° and 4.1° for films on Fe(001) and Fe(110) respectively.

Concluding Remarks

In summary the passive film that forms on pure iron at a high potential in near neutral solution has been definitively shown to be crystalline with a highly defective nanophase structure. These results open up the way for detailed modelling of the films propoerties and to reaching a fundamental understanding of protection and breakdown in engineering systems. This work also provides a point to address more complex issues in passive film studies such as kinetic and environemtal effects during film growth.

Finally, the most common engineering approaches to corrosion protection involve adding alloying elements to the metal (stainless steel) or inhibiting chemicals to aggressive environments. Our results on iron provide a starting point for understanding how these approaches affect passive films and provide enhanced corrosion protection.

Acknowledgements

We thank Jean Jordan-Sweet for assistance with X20A, Chi-Chang Kao for the generous use of his detector, Carissima Vitus for early experimental work, and Ruediger Dieckmann for useful electronic discussions. This work was performed partly under the auspices of the US DOE, Division of Materials Science, Office of Basic Energy Sciences, under contract DE-AC02-76CH00016.

References

1. Keir, J. *Phil. Trans. Royal. Soc* 1790, *80*, 359.
2. Faraday, M. *Experimental Researches in Electricity*; Dover: New York, 1965; Vol. 2.
3. Uhlig, H. H. ; Frankenthal, R. P. and Kruger, J., Ed.; The Electrochemical Society: Princeton NJ, 1983, pp 1.
4. Evans, U. R. *J. Chem. Soc* 1927, *127*, 1020.
5. Iitaka, I.; Miyake, S.; Iimori, T. *Nature* 1937, *139*, 156.
6. Markovac, V.; Cohen, M. *J. Electrochem. Soc.* 1967, *114*, 678.
7. Nagayama, M.; Cohen, M. *J. Electrochem. Soc* 1963, *110*, 670.
8. O'Grady, W. E. *J. Electrochem. Soc* 1980, *127*, 555.
9. Eldridge, J.; Hoffman, R. W. *J. Electrochem. Soc* 1989, *135*, 955.
10. Kerkar, M.; J. Robinson; Forty, A. J. *Faraday Discuss. Chem. Soc* 1990, *89*, 31.
11. Long, G. G.; Kruger, J.; D. R. Black; Kuriyama, M. 1983, *150*, 603.
12. Rubim, J. C.; Dünnwald, J. *J. Electroanal. Chem* 1989, *258*, 327.
13. Gui, J.; T.M. Devine, (1991) p1105. *Corrosion Sci.* 1991, *32*, 1105.
14. Davenport, A. J.; Sansone, M. *Journal of the Electrochemical Society* 1995, *142*, 725.
15. Ryan, M. P.; Newman, R. C.; Thompson, G. E. *J. Electrochem. Soc* 1995, *142*, L177.
16. Long, G. G.; Kruger, J.; D. R. Black; Kuriyama, M. *J. Electrochem. Soc.* 1983, *130*, 240.
17. Kirchheim, R. *Corrosion Science* 1989, *29*, 183.
18. Urquidi, M.; MacDonald, D. D. *J. Electrochem. Soc.* 1985, *132*, 555.
19. Toney, M. F.; Gordon, J. G.; Samant, M. G.; Borges, G. L.; Melroy, O. R.; Kau, L. S.; Wiesler, D. G.; Yee, D.; Sorensen, L. B. *Phys. Rev. B* 1990, *42*, 5594.
20. Oblonsky, L. J.; A. J. Davenport; Ryan, M. P.; Isaacs, H. S.; Newman, R. C. *J. Electrochem. Soc* 1997, *144*, 2398.
21. Toney, M. F.; Davenport, A. J.; Oblonsky, L. J.; Ryan, M. P.; Vitus, C. M. *Physical Review Letters* 1997, *79*, 4282.
22. Wyckoff, R. W. *Crystal Structures*; Interscience Publishers: New York, 1965; Vol. 3.
23. Dieckmann, R. *Solid State Ionics* 1984, *12*, 1.
24. McHale, J. M.; Auroux, A.; Perrotta, A. J.; Navrotsky, A. *Science* 1997, *277*, 788.
25. Siegel, R. W. *J. Phys. Chem. Solids* 1994, *55*, 1097.
26. Kuroda, K.; Cahan, B. D.; Nazri, G.; Yeager, E.; Mitchell, T. E. *J. Electrochem. Soc* 1982, *129*, 2163.
27. Sewell, P. B.; Stockbridge, C. D.; Cohen, M. *J. Electrochem. Soc,* 1961, *108*, 933.

Chapter 8

Modeling Dynamic Properties of Mineral Surfaces

N. H. de Leeuw, S. E. Redfern, D. J. Cooke, D. J. Osguthorpe, and S. C. Parker[1]

Department of Chemistry, University of Bath, Claverton Down, Bath BA2 7AY, United Kingdom
[1]Corresponding author: s.c.parker@bath.ac.uk

Abstract

We describe the application of atomistic simulation techniques to investigate mineral surface properties, such as the effect water adsorption on the structure and stability of baryte ($BaSO_4$) surfaces, the energetics of calcite ($CaCO_3$) growth at surface steps and the determination of the surface vibrational properties of Fe_2O_3,. Hydration stabilises all baryte surfaces, although the {210} surface remains the dominant surface in agreement with experiment. The interaction of water molecules with both calcite and baryte surfaces is predominantly through coordination of the water molecule's oxygen atom to the surface cation, rather than hydrogen-bonding to surface oxygen atoms. On baryte the hydration energies (-53 to -66 $kJmol^{-1}$) indicate physisorption of the water molecules, while on calcite the energies (-79 to -139 $kJmol^{-1}$) are more in the region of chemisorption. The rate determining step of calcite growth from two step edges is the creation of the kink sites on the edges when the first $CaCO_3$ or $MgCO_3$ unit is introduced. Pure calcite growth is calculated to occur preferentially from the obtuse step edge, where introduction of the kink sites costs 82.0 $kJmol^{-1}$ as opposed to 235.4 $kJmol^{-1}$ on the acute edge. Addition of $MgCO_3$ to both purely calcite edges and edges decorated by magnesium is an exothermic process and at all stages energetically more favourable than addition of $CaCO_3$ units. Hence in a mixed solution magnesium will be preferentially incorporated into the growing calcite crystal and pure calcite growth is thus inhibited. Finally, simulations on hematite surfaces suggest that the surface free-energy is dominated by the potential and zero point energies and that the contribution of vibrational entropy is minimal.

[1] Corresponding author, email: s.c.parker@bath.ac.uk

INTRODUCTION

The aim of our work is to model surface structure, reactivity and dynamical processes of minerals at the atomic level. In this paper we describe our progress on a number of minerals, namely hematite (Fe_2O_3), calcite ($CaCO_3$) and baryte ($BaSO_4$). In earlier work [1-9] we have shown that simulation techniques can model the interaction of water with perfect surfaces. We are now attempting to extend this in three ways, first, by studying the energetics of adsorption at different surfaces and surface defects that are likely to be present on real crystals, as illustrated by the planar and stepped surfaces of baryte. Secondly, by modelling the energetics of the growth of calcite and the effect of incorporating magnesium ions in aqueous solution. An additional aim is to investigate the ability of the solvent to modify the energetics of the different stages of the growth process. Thirdly, by investigating ways of calculating the surface free-energy which ultimately governs the stability and reactivity of surfaces.

At present atomistic simulation techniques represent the best way of rapidly modelling many different surfaces of different materials. The basis of this approach is the Born model of solids where simple parameterised analytical equations are used to describe the forces between atoms. Although a full electronic structure calculation is preferred the inherent speed of atomistic simulations allows us to examine many different models of different surfaces with a high level of complexity. The results can then be used as an aid in the interpretation of experiment and thus provide a useful complement to experimental structural techniques, provided of course that the methods are reliable. There are now many examples in the literature where there is good agreement between the structural models developed by the atomistic simulation methods and those observed by experiment. Such examples include the surfaces of rutile [10], alumina [11] and tungsten oxide [12]. In addition to giving good structural information these techniques also provide detailed energetics and where there are available experimental data the energies of adsorption of water, for example on magnesia, calcia [2] and alumina [8,9] are in good agreement. The level of agreement between simulation and experiment gives us sufficient confidence to apply these techniques to problems where there are much less experimental data. We will describe our recent progress in three such areas. However, before discussing applications we will briefly review the methods.

THEORETICAL METHODS

The atomistic simulation of the surfaces of polar solids was pioneered by the work of Tasker [13,14] and Mackrodt and Stewart [11]. The low energy, and hence the most common, surfaces of a crystal are generally those of low Miller index. These planes are the closest packed and hence have the smallest surface area per unit cell. As the unit cell volume is fixed these will have the largest interplanar spacings, they will then be most easily cleaved and hence have the lowest surface energy. However in polar solids other factors apply, for example the electrostatic contribution. Bertaut

[15] demonstrated that when there is a dipole moment perpendicular to the surface, the surface energy diverges and increases with increasing size. The surface must therefore be constructed with no net dipole perpendicular to the surface [14]. Thus surfaces which are made of alternating charged planes, called dipolar (or polar) surfaces cannot occur naturally without the adsorption of foreign atoms or surface roughening [16]. As a way of identifying these potentially unstable surfaces Tasker [13] characterized the surfaces in terms of the repeat unit which when repeated into the bulk generates the crystal. He identified three types of surface, type I in which the repeat unit is charge neutral stoichiometric layers, type II which is comprised of charged layers but in such a way that there is no dipole moment perpendicular to the surface and finally type III where there is such a dipole moment, which has to be modified in some way to remove the dipole.

We use two strategies for generating a simulation cell for modelling free surfaces. One stratergy uses a two region approach and 2-dimensional periodicity is assumed. The crystal is divided into two regions; a region I adjacent to the interface where the ions are allowed to move independently and a region II in which the ions are held fixed relative to each other but the region as a whole may move. The advantage of using a two region approach is that inclusion of a region II ensures that all the energies of the ions in region I are fully converged. The second stratergy is to take only region I and assume three-dimensional periodicity. Thus the simulation cell comprises a slab of solid separated by a vacuum gap or fluid which is then repeated infinitely. The virtue of this approach is that it exploits the periodicity to ensure that the energies, particularly the electrostatic component, of large simulation cells can be calculated rapidly using efficient algorithms but the disadvantage is that as a consequence of the long range nature of the electrostatic forces the ions on one surface may be influenced by the behaviour of ions on the other.

Once the energies and forces are evaluated we next apply either energy minimisation or molecular dynamics simulation techniques. Energy minimisation is achieved by adjusting the atom positions in region I until the net forces on each atom are zero. Molecular dynamics allows explicit treatment of temperature by giving all the ions in region I kinetic energy. Generally, we begin by using energy minimisation to evaluate the surface structure and energy of a range of surfaces and select suitable candidate surfaces for further study with molecular dynamics. Again we first investigate the surface structure and energy before considering dynamical properties, such as molecular transport.

The potential parameters used for the simulation of the calcite crystal are those derived by Pavese et al. [17] in their study of the thermal dependence of structural and elastic properties of calcite while the potential model used for baryte was derived by Allen et al. [18]. The inter- and intramolecular interactions of the water molecules were derived by ourselves [5]. The calcite/water interactions were verified against the structure of the calcium carbonate hexahydrate ikaite as described previously [5], and the close agreement between experimental and calculated structure and formation energies means that we may be confident that the potential parameters describe the water/surface interactions adequately. When considering impurity ions, e.g. magnesium, our view is that the relative hydration energies, e.g. between calcium and

magnesium ions, should agree with experiment and the relative energies are included in the fitting. The interatomic potentials for hematite were those derived by Lewis and Catlow [19].
In the next section we describe our recent work on modelling the interaction between water and mineral surfaces, our model for simulation of the energetics of crystal growth and finally, preliminary work on modelling surface free-energies.

RESULTS

The interaction of water with mineral surfaces

Calcite ($CaCO_3$), the predominant polymorph of calcium carbonate and one of the most abundant minerals, and baryte ($BaSO_4$) co-exist in the marine environment and often co-precipitate. Calcite is important as a building block of shells and skeletons [20,21], in ion exchange [22]and together with baryte as a scale former [23] and as such has been the subject of extensive research [e.g. *24-30*]. We used energy minimisation techniques to model the hydration of a range of calcite and baryte surfaces by associatively adsorbing up to one water molecule per surface cation. The initial sites chosen for the adsorption of water onto the surfaces were above the surface cations and oxygen atoms. The reason we model one monolayer of water in static calculations is three-fold. Firstly, the results, such as hydration energies, are directly comparable to most surface science techniques, including temperature programmed desorption (TPD) and microcalorimetric measurements, which mainly use materials generated under ultra-high vacuum conditions dosed with gaseous water (see for example Fubini et al. [*31*]). Secondly, in static calculations more than one monolayer of water would be modelling the mineral/ice interface rather than liquid water giving interfacial rather than surface energies and properties. Thirdly, molecular dynamics simulations of MgO in liquid water showed that there is a distinct difference in structure and density between the adsorbed monolayer and the bulk water, and the static calculations with one adsorbed monolayer mimick these MD simulations [6].

Calcite

Calcite has a rhombohedral crystal structure with spacegroup $R\bar{3}c$ and a = b = 4.990 Å, c = 17.061 Å, $\alpha = \beta = 90^0$ and $\gamma = 120^0$ [*32*] which on energy minimisation, using the potential model developed by Pavese et al. [*17*], as calculated to be a = b = 4.797 Å, c = 17.482 Å, $\alpha = \beta = 90^0$ and $\gamma = 120^0$. The surface structure and stability of the hydrated calcite surfaces have been extensively discussed in previous work [*1,4,5*] and we therefore briefly summarise the salient points. The relevant surface and hydration energies are collected in Table 1. For an explanation of the procedure for calculating the hydration energies see Parker et al. [*1*].

Table 1. Surface and Hydration Energies of Calcite

Surface and Hydration Energies of Calcite			
Surface (hexagonal indices)[*]	γ_{pure} / Jm^{-2}	γ_{hydrated} / Jm^{-2}	$E_{\text{hydration}}$ / kJmol^{-1} (H_2O)
$\{10\bar{1}4\}$	0.59	0.16	- 93.9
$\{1001\}$ Ca	0.97	0.68	- 79.2
$\{1001\}$ CO$_3$	0.99	0.38	- 93.2
$\{10\bar{1}0\}$	0.97	0.75	-100.5
$\{10\bar{1}1\}$ Ca	1.23	0.63	-113.4
$\{10\bar{1}1\}$ CO$_3$	1.14	0.81	-100.9
$\{10\bar{2}0\}$	1.39	0.43	-138.5

*The element following the index indicates the top most atoms for a polar surface

From the surface energies in Table 1. it is clear that the $\{10\bar{1}4\}$ surface is by far the most stable surface in the calcite crystal and it is found to dominate the morphology [*26, 33-37*]. The surface consists of layers containing both calcium ions and carbonate groups. The surface plane is always terminated with oxygen atoms. The layers contain two differently oriented carbonate groups and as a result the oxygen atoms on the surface are in two different but equivalent orientations. On hydration, the water molecules adsorb uniformly onto the $\{10\bar{1}4\}$ surface in a herringbone pattern. The water molecules' oxygen atoms are coordinated to the surface calcium ions at a distance of 2.4 Å, with the two hydrogen atoms pointing towards two surface oxygen ions. On the fully hydrated surface, all surface oxygen atoms are coordinated to two hydrogen atoms from two different water molecules. Although the hydrogen-bond lengths are different for the two water molecules (1.89 and 1.97 Å) the different orientations are still equivalent, [*4*]. The $\{10\bar{1}4\}$ surface exhibits 1 x 1 symmetry in accordance with the AFM images of the $\{10\bar{1}4\}$ plane of calcite in water by Ohnesorge and Binnig [*38*]. The surface energy of the hydrated $\{10\bar{1}4\}$ surface of γ = 0.16 Jm^{-2} agrees well with the experimental surface energy of 0.23 Jm^{-2} found by Gilman [*39*] obtained by cleaving the crystal, particularly when taking into account that the experimental surface was mechanically cut and will contain steps and other dislocations which will raise the surface energy.

Generally, any carbonate groups that are in an upright position on a surface, rotate upon relaxation to lie flat in the surface. This effect is strongest when the surfaces are half vacant in carbonate groups (necessary to remove any dipole moment perpendicular to the surface). The surface cations do not show any significant

102

relaxation. Upon hydration, the water molecules preferentially adsorb to surface cations. There is usually little interaction between the adsorbed water molecules (O—H distance > 2.4 Å). In contract, the carbonate terminated {0001} surface shows extensive intermolecular coordination between the adsorbed water molecules which accounts for an extra stabilising energy factor of -1.5 kJmol^{-1}. However, this is negligible compared to the hydration energy of -93.2 kJmol^{-1} which is thus due almost solely to the interactions between water molecules and surface ions. All calcite surfaces are stabilised by the adsorption of water molecules. The hydration energies for the various surfaces fall in a fairly narrow range of -79.2 to -138.5 kJmol^{-1} which at least for the more reactive surfaces can probably be classed as chemisorption of water. The hydration energies of the planar and stepped surfaces are the same indicating that the water molecules show no preference to adsorb at the edges. This is probably due to the very regular pattern of adsorption on the planar surface which is disturbed by the steps.

Baryte

Baryte ($BaSO_4$) has an orthorhombic crystal structure with spacegroup Pnma. Its lattice parameters are a = 8.88 Å, b = 5.45 Å, c = 7.15 Å, $\alpha = \beta = \gamma = 90^0$ [32] which relaxed on energy minimisation, employing the potential model by Allan et al. [18] to a = 8.88 Å, b = 5.47 Å, c = 7.14 Å, $\alpha = \beta = \gamma = 90^0$. The morphology is dominated by the {210} and {001} faces [40,41] and some impurities (especially Sr and some Pb and Ca) are often present in the crystal [32,42]. In addition to the {001} and {210} cleavage planes, we considered the {100} and {410} surfaces. The {410} surface contains {210} planes separated by monatomic steps consisting of {100} step walls. Baryte crystals grow in layers from steps [30] and as such we investigated the {410} surface as a model of a stepped {210} surface.

All four baryte planes contain a stoichiometric ratio of both barium atoms and sulfate groups in each layer including the surface layer and thus there is no dipole moment perpendicular to the surfaces. The sulfate groups, which are a tetrahedral arrangement of four oxygen atoms surrounding the sulfur atom, may be positioned in the surface either edge uppermost with two oxygen atoms at the surface as on the {001} plane, or alternatively face uppermost with three oxygen atoms at the surface, as for example in the {210} plane, where after energy minimisation the edge is somewhat rotated exposing a third oxygen atom at the surface. The unhydrated {210} and {001} cleavage planes have the lowest surface energies and are hence the most stable planes (Table 2). This is in agreement with work of Allan et al. [18] and Parker et al. [36]. Apart from some rotation of the sulfate groups on the {210} surface, neither plane shows any noticcable relaxation. The {410} surface containing {210} planes separated by steps is as expected somewhat less stable than the perfect {210} surface. The steps are approximately 10 Å apart, which makes interactions between the steps unlikely.

The {100} is the least stable of the four surfaces studied and can be cut in two ways giving similar surface energies (Table 2). One termination is a smooth plane giving the lower surface energy of 0.91 Jm^{-2}. It consists of sulfate groups which are both face exposed and edge exposed. The latter rotate slightly on relaxation to become more face exposed and also move into the bulk crystal together with the surface barium atoms which recede into the bulk by about 0.3 Å. The other possible cut is a micro-facetted surface whereby alternate rows of $BaSO_4$ units are removed, leading to parallel edges containing both barium atoms and edge exposed SO_4 groups, which rotate on relaxation to be face on to the surface. Although these micro-facetted planes were very unfavourable in the case of the {210} and {001} surfaces, the surface energy of the micro-facetted {100} plane at 0.93 Jm^{-2} is very similar to the planar {100} surface and thus we need to consider both. This situation, where a relatively unstable surface is stabilised, or at least not destabilised, by micro-facetting is similar to rocksalt structured MgO, where the surface energies of the relatively unstable {110} and {111} surfaces were drastically lowered upon micro-facetting of the surfaces which resulted in the expression of planes of the dominant {100} surface separated by edges and corners [2,3].

In general, only the sulfate groups show significant relaxation after energy minimisation of the surfaces. They rotate in the surface similar to the behaviour of surface carbonate groups in calcium carbonates, especially aragonite [5].

Table 2. Surface and Hydration Energies of Baryte

Surface and Hydration Energies of Baryte			
Surface	γ_{pure} / Jm^{-2}	$\gamma_{hydrated}$ / Jm^{-2}	$E_{hydration}$ / $kJmol^{-1}$ (H_2O)
{210}	0.55	0.49	-52.6
{410}	0.73	0.54	-61.8
{410}[a]	0.73	-	-66.0
{001}	0.59	0.51	-55.0
{100}[b]	0.91	0.85	-53.1
{100}[c]	0.93	0.79	-56.9

[a] adsorption on edges only, [b] planar surface, [c] micro-facetted surface, with alternate rows of BaSO4 units removed

Adsorption of water at the baryte {001} surface has hardly any effect on the surface structure which remains almost bulk-terminated. Some of the surface sulfate groups have smaller O-S-O angles (110.5-111.5^0) than on the unhydrated surface (111.0-112.5^0) due to hydrogen-bonding of the sulfate oxygen atoms to the water molecules' hydrogen atoms. The surface energy has decreased as a result of the adsorbed water molecules. The hydrated {210} surface remains the most stable surface considered (Table 2) with the lowest surface energy. The effect of the monolayer of water molecules adsorbed on the surface is to anchor the surface sulfate groups in a bulk-like position reducing the extent of rotation which was found on the relaxed unhydrated surface. The O-S-O angles in the surface sulfate groups, which were a uniform 111.5^0 on the unhydrated surface, now vary between 110.5-111.8^0, again due to hydrogen-bonding to the water molecules.

As mentioned above, we studied the {410} surface as a model for a stepped {210} surface. Hence we not only studied a full monolayer of water on the surface but also investigated whether the edge sites were the more reactive sites for adsorption of water. For example, Eggleston et al. [43] used Fourier transform infrared spectroscopy and scanning tunneling microscopy to show that aqueous sulfate species preferentially adsorb at hematite step sites, while previous work by ourselves [3] has shown that on stepped MgO surfaces, the preferred adsorption sites are at the step edges rather than on the planar surface due to the lower coordination (4-coordinated instead of 5-coordinated) of the edge species. Hydration along the edges of the baryte {410} surface shows the same effect. The hydration energy for adsorption at the edges of the step of -66.0 kJmol^{-1} (Table 2) is larger than the hydration energy for full monolayer coverage (-61.8 kJmol^{-1}), where the water molecules are evenly spaced over the entire surface. This indicates that adsorption preferentially occurs on step edges of the {410} surface rather than on the {210} planes, in agreement with our earlier findings on MgO. On hydration of the planar and micro-facetted {100} surfaces the relative stabilities of the two planes are reversed. The micro-facetted plane is stabilised by the adsorption of water to a larger extent than the planar surface and has a lower surface energy. This indicates that in a wet environment, the micro-facetted {100} surface should be found to prdominate, even though the surface energies of the two hydrated planes are similar.

The hydration energies of the four hydrated baryte surfaces are very similar and range from -52.6 kJmol^{-1} on the dominant {210} surface to -66.0 kJmol^{-1} at the step edges of the {410} plane. Contrary to the hydration energies of the calcite surfaces which are both larger and range more widely, the hydration energies of the baryte surfaces indicate physisorption of water molecules onto the surfaces. This reduced interaction of the water molecules with the surface species is indicated by a smaller stabilisation of the surfaces compared to the much larger decrease in surface energies of the hydrated calcite surfaces. For example, the surface energy of the calcite {10$\bar{1}$4} surface has decreased by 73% but only by 11% for the baryte {210} surface.

However, the two minerals correspond in their adsorption behaviour in that the interaction between surface cation and the oxygen atom of the water molecule is the most important, with hydrogen bonding to surface oxygen atoms a significant but secondary effect. The water molecules themselves usually do not interact significantly with each other, indicating that adsorption to the surface outweighs possible intermolecular interactions between the water molecules, such as hydrogen-bonding. Hydrogen-bonding to both surface oxygen atoms and intermolecularly to other water molecules is found for both systems, although in the calcium carbonate crystals hydrogen-bonds to the lattice oxygen atoms are usually shorter (1.5-2.0 Å) than in baryte (2.1-2.5 Å). The situation where the water molecule adsorbs by only its two hydrogen atoms to two surface oxygen atoms is not found in calcium carbonate, possibly due to the larger O-O distance in the calcium carbonate (~2.0 Å) and the larger and more rigid O-C-O angle (120^0) which makes it more difficult for the water molecules' hydrogen atoms to coordinate to the carbonate oxygen atoms.

Models for Crystal Growth

In addition to modelling the mineral surfaces we are interested in developing strategies for modelling crystal growth, particularly the role that impurities and additives play in modifying the growth rates. At low and medium supersaturation levels crystal growth is found to occur from steps [24] or spiral dislocations [25] depending on the supersaturation, and often as observed by Liang et al. [26] monolayers form from steps. Recent models of step dissolution have included a terrace-ledge-kink model, successfully describing the initial stages of pit growth on the surface [27,28], and a kinetic Monte Carlo model which reproduces experimental pit-growth behaviour [29]. Atomistic simulation methods have been used to model growth inhibition by incorporation of diphosphates into the steps [44] although they did not explicitly include solvent effects. Crystal growth is often affected by the incorporation in the crystal of foreign ions such as cadmium or magnesium, e.g. Compton and Brown [45] found that magnesium ions inhibit calcite growth. In this section we describe how we employed Molecular Dynamics (MD) simulations to investigate calcite growth at steps in aqueous environment, both as pure calcite and in the presence of magnesium. We studied the vicinal $\{31\,\bar{4}\,8\}$ and $\{3\,\bar{1}\,\bar{2}\,16\}$ surfaces, which are made up of $\{10\,\bar{1}\,4\}$ planes separated by monatomic steps (Figure 1). The walls of the steps are also $\{10\,\bar{1}\,4\}$ planes and the step edges of the two stepped surfaces are identical to the two different edges of the calcite crystal which has rhombohedral morphology [36]. The steps on the $\{31\,\bar{4}\,8\}$ surface are acute, *i.e.* the carbonate group on the edge of the step overhangs the plane below the step (Figure 1a) and the angle between step wall and plane is 80^0 on the relaxed surface (cf. 78^0 exp. [22]). The steps on the $\{3\,\bar{1}\,\bar{2}\,16\}$ surface on the other hand are obtuse, *i.e.* the carbonate groups on the step edge are leaning back with respect to the plane below (Figure 1b) with an angle between step wall and plane of 105^0 on the relaxed surface (exp. 102^0 [22]). These two types of step are found experimentally to form the dissolving edges of etch pits [22,26] the obtuse step is found to be the fastest moving of the two [46]. The surfaces were modelled as a repeating crystal slab, containing 96 $CaCO_3$ units, and a void and run using the NVT ensemble. Water molecules were then introduced in the void and the whole system including 48 solvent molecules was again simulated under NVT conditions. We used the Nose-Hoover algorithm within the DLPOLY code [47] for the thermostat with parameters set at 0.5 for both the thermostat and barostat relaxation times (ps). The oxygen shell mass was set at 0.2 a.u., which is small compared to the mass of the hydrogen atom of 1.0 a.u. to ensure that there would be no exchange of energy between vibrations of oxygen core and shell with oxygen and hydrogen vibrations [6]. However, due to the small shell mass we needed to run the MD simulation with a small timestep of 0.2 femtoseconds.

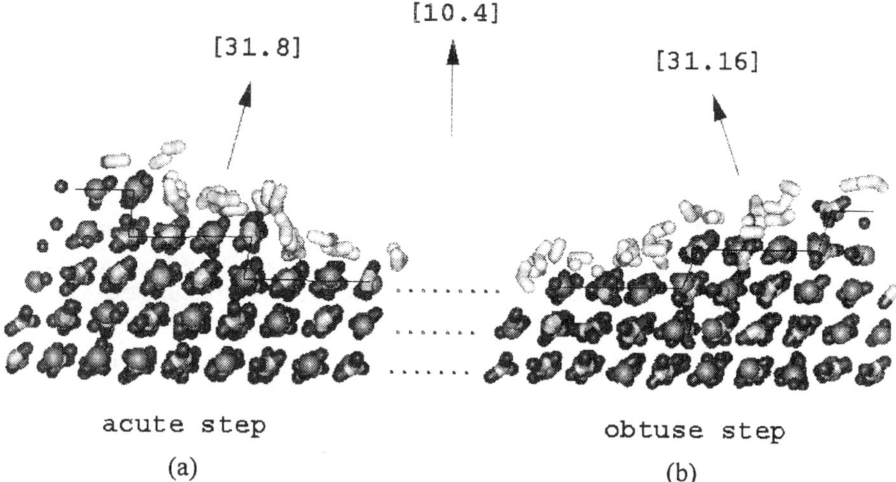

Figure 1 A snap shot of the vicinal surfaces used for modelling step growth showing (a) the acute step modelled by {31 $\bar{4}$ 8} and (b) the {3 $\bar{1}$ $\bar{2}$ 16} the obtuse step and a layer of water molecules.

At 300K, the water molecules adsorb onto the planar calcite surface in the same regular herringbone pattern as observed using static techniques. As expected, with the introduction of temperature the hydration energy decreased to -69.0 kJmol^{-1}, compared to -93.9 kJmol^{-1} from the static calculations (Table 1). Once adsorbed the water molecules stay on the surface and do not diffuse through the gap at the simulated temperature (300K). We modelled the stepped surfaces under the same conditions and calculated the energies of adding a succession of calcium carbonate units to the steps, mimicking growth under saturated conditions. In effect, we have simulated and obtained an energy for the reaction given in the following equation:

$$nCaCO_{3(s)} + Ca^{2+}_{(aq)} + CO^{2-}_{3(aq)} \rightarrow (n+1)CaCO_{3(s)}$$

We then performed a series of simulations where we added magnesium carbonate units to the calcite step edges and subsequently MgCO$_3$ and CaCO$_3$ units to the magnesium edges. The energies are collected in Figure 2 where each 25% growth corresponds to one unit added to the growing step. Figure 2(a) shows that addition of the initial CaCO$_3$ unit to the calcite step, which introduces kink sites on the step edge, is much more endothermic at the acute edge (+235.4 kJmol^{-1}) than at the obtuse step (+82.0 kJmol^{-1}). The energies of adding subsequent units are less endothermic than the first unit and below 100 kJmol^{-1} for both step edges.

Thus, introduction of the kink sites is the rate limiting step of the growth process and on thermodynamic grounds growth at the obtuse step should occur. Figure 2(a) also shows that addition of MgCO$_3$ units is an exothermic process at all stages at both step edges and hence, in the presence of magnesium ions, growth of MgCO$_3$ should occur preferentially at both step edges rather than CaCO$_3$ growth, incorporating magnesium into the calcite crystal. In Figure 2(b) we compare the growth energetics on steps decorated with magnesium ions. This time the initial growth of a $CaCO_3$ unit onto the acute step edge is energetically more favourable than growth onto the obtuse step. However, once again $MgCO_3$ growth onto the edges is exothermic with similar energies for the two steps. The results therefore suggest that incorporation of magnesium ions into the growing calcite crystal is energetically favourable and the presence of magnesium ions inhibits calcite growth.

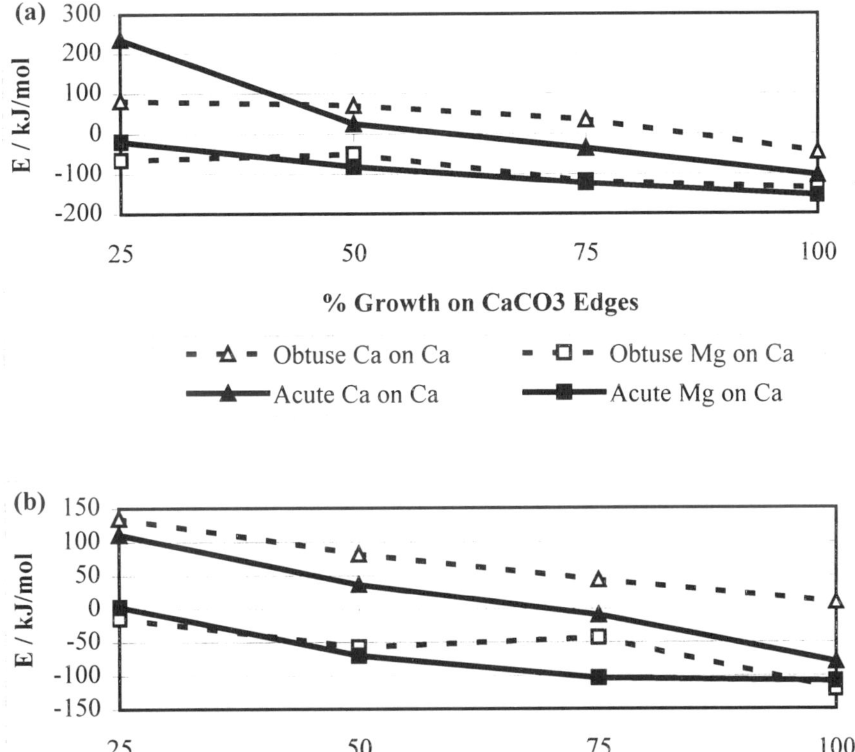

Figure 2. Graphs showing CaCO3 and MgCO3 growth on obtuse and acute calcite edges (a) growth onto a calcium edge and (b) growth onto magnesium decorated edge.

Modelling the Vibrational Density of States

The most efficient method for calculating the vibrational frequencies and the free-energy of solids is the lattice dynamics approach [49]. The approach assumes that the vibrational motions are quantised harmonic oscillators depending only on the cell volume, which is called the quasi-harmonic approximation. The vibrational frequencies are then simply calculated by diagonalising the dynamical matrix and then using simple statistical mechanical expressions to give the thermodynamic properties including free-energy. We have followed this method and have successfully modelled the structure and thermodynamic properties of bulk minerals [50]. The approach depends for its success on the assumption that the anharmonic contribution to the free-energy is dominated by the extrinsic component, i.e. vibrational frequencies vary with cell dimension, and hence intrinsic anharmonicity, which relates to the interaction between the modes, is negligible. When considering the equations of state and thermodynamics of bulk minerals at high pressure this holds. However, at low pressures, such as occurs at free surfaces and at high temperatures alternative strategies need to be considered. The technique which implicitly includes intrinsic anharmonicity is molecular dynamics. Thus we have used MD coupled with an analysis code FOCUS [51] to calculate the surface free-energies. One of the program FOCUS's features is that it extracts information equivalent to density of states spectra and normal modes from molecular dynamics trajectories using digital signal processing techniques. The approach is to take the coordinate (or velocity) trajectory of each atom and Fourier transforming to the frequency domain, using the discrete fourier transform, and after appropriately weighting and converting the frequencies to the phonon density of states (DOS).

We have begun by modelling the most stable low index surfaces of hematite, Fe_2O_3, namely $\{0001\}$ and $\{10\bar{1}2\}$ [52] using well-established interatomic potentials [19]. Figure 3 shows the calculated density of states for (a) the bulk crystal and slabs containing; (b) an Fe terminated $\{0001\}$ crystal (c) and (d) with the Fe and O terminated $\{10\bar{1}2\}$ surfaces respectively. The approximate size of the simulation cells are 8 nm^3 and because the potential model includes shell model to give a representation of electronic polarisability the time step is 0.1 fs. Furthermore, to ensure resolution better than 1 cm^{-1} and data on frequencies above 4000 cm^{-1} the simulations must typically run for greater than 1 ns and full coordinate and velocity data collected every 2 fs. Although the phonon DOS in Figure 3 are qualitatively similar there are some marked differences, for example in Figure 3b the $\{0001\}$ shows significant low frequency modes which are absent from the $\{10\bar{1}2\}$ surfaces. The calculated surface energies, obtained from the potential energy, are 2.5, 2.8 and 3.0 Jm^{-2} for the $\{0001\}$Fe, $\{10\bar{1}2\}$Fe and $\{10\bar{1}2\}$O surfaces respectively. On evaluating the surface free-energies at 300K the Fe terminated surfaces remain more stable but the differences are much smaller i.e. they are 3.1, 3.1 and 3.2 Jm^{-2} for the $\{0001\}$Fe, $\{10\bar{1}2\}$Fe and $\{10\bar{1}2\}$O surfaces respectively. Perhaps surprisingly the differences between the surface energy and surface free-energy are not due to the vibrational entropy but to the zero point energies which accounts for about 94% of the difference.

(a)   (b)

(c)   (d)

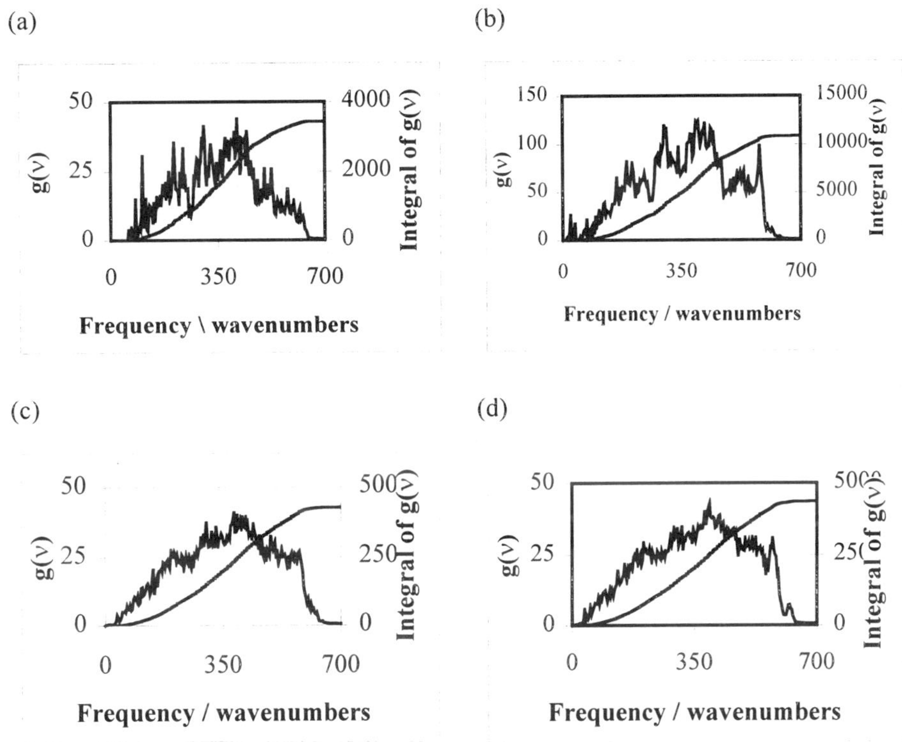

Figure 3 Phonon density of states for (a) the bulk crystal, (b) a slab with the Fe terminated {0001} surface (c) the Fe terminated {10$\bar{1}$2} surface and (d) the O terminated {10$\bar{1}$2}

CONCLUSION

We have employed atomistic simulations, using both energy minimisation techniques and molecular dynamics simulations, to investigate surface properties such as vibrational modes to determine surface free energies, the effects of surface hydration and crystal growth . As a result, we can make the following observations:

Simulation of the hydration of both calcite and baryte surfaces show that adsorption of water has a stabilising effect with the calcite {10$\bar{1}$4} and baryte {210} surfaces remaining the dominant crystal planes. The mode of adsorption is predominantly through coordination of the water molecule's oxygen atom to surface cations. The hydration energies are determined by this interaction with a very minor contribution from intermolecular hydrogen-bonding between the water molecules. On baryte the hydration energies (-53 to -66 kJmol^{-1}) indicate physisorption of the water molecules, while on calcite the energies (-79 to -139 kJmol^{-1}) are more in the region of chemisorption.

Molecular dynamics simulation of $MgCO_3$ and $CaCO_3$ growth onto acute and obtuse edges on the calcite $\{10\bar{1}4\}$ surface shows that calcite growth occurs preferentially at the obtuse steps, in agreement with experimental findings, where the rate limiting step is introduction of the kink sites. Addition of $MgCO_3$ to both calcite steps is an exothermic process and hence occurs in preference to pure calcite growth. Subsequent growth onto the edges decorated by magnesium ions is again exothermic for $MgCO_3$ but not $CaCO_3$. Hence calcite growth is inhibited in the presence of magnesium ions.

Although preliminary our results on Fe_2O_3 suggest that for a solid surface the entropic contribution to the free-energy is small and that other than the potential energy the zero point energy is the only significant term.

The results presented above show that atomistic simulation techniques are a viable tool for modelling the structure, stability and coordination of mineral surfaces, including dynamical processes. However, they do depend on reliable interatomic potentials and hence there is a great need for both experiments which can measure atomic proceses and electronic structure calculations on small systems to compare and test the reliability of the models.

Acknowledgement

We thank EPSRC and NERC for financial support.

References

1. Parker S.C., de Leeuw N.H. and Redfern S.E. *Faraday Discussions*, **1999**, 114, 381-389

2. de Leeuw N.H.; Watson G.W.; Parker S.C. *J. Phys. Chem.* **1995**, *99*, 17219-17225.

3. de Leeuw N.H.; Watson G.W.; Parker S.C. *J. Chem. Soc. Faraday Trans.* **1996**, *92*, 2081-2091.

4. de Leeuw N.H.; Parker S.C. *J. Chem. Soc. Faraday Trans.*, **1997**, *93*, 467.

5. de Leeuw N.H.; Parker S.C. *J. Phys. Chem. B* **1998**, *102*, 2914-2922.

6. de Leeuw N.H.; Parker S.C. *Phys. Rev. B* **1998**, *58*, 13901-13908.

7. de Leeuw N.H.; Higgins F.M.; Parker S.C. *J. Phys. Chem. B* **1998**, *103*, 1270-1277.

8. de Leeuw N.H.; Parker S.C. *J. Am. Ceram. Soc.* **1999**, 82, 3209-3216.

9. de Leeuw N.H.; Redfern S.E.; Parker S.C. *Recent Res. Devel. Phys. Chem.*, **1998**, 2, 441.

10. Oliver P.M.; Watson G.W.; Kelsey E.T.; Parker S.C., *J. Mater. Chem.* **1997**, 7, 563.

11. Mackrodt W.C.; Stewart R.F *J. Phys. C* **1979**, 12, 431-449.

12. Oliver P.M.; Parker S.C.; Egdell R.G.; F.H. Jones F.H. *J. Chem. Soc. Faraday Trans.* **1996**, 92, 2049.

13. Tasker P.W. *J. Phys. C: Solid State Phys.* **1979**, 12, 4977-4983.

14. Tasker P.W. *Phil. Mag. A* **1979**, 39, 119-136.

15. Bertaut F. *Compt. Rendu.*, **1958**, 246, 3447.

16. Benson G.L.; Yon K.S. In *The Solid Gas Interface*; Flood E.A., Ed., Arnold, London, UK, 1967

17. Pavese A.; Catti M.; Parker S.C.; Wall A. *Phys. Chem. Miner.* **1996**, 23, 89-93.

18. Allan N.L.; Rohl A.L.; Gay D.H.; Catlow C.R.A.; Davey R.J.; Mackrodt W.C. *Faraday Discuss.* **1993**, 95, 273-280.

19. Lewis G.V.; Catlow C.R.A., *J. Phys. C: Solid State Phys.*, **1985**, 18, 1149

20. Beruto D.; Giordani M. *J. Chem. Soc. Faraday Trans.* **1993**, 89, 2457-2461.

21. Romanek C.S.; Grossman E.L.;Morse J.W. *Geochim. Cosmochim. Acta* **1992**, 56, 419.

22. Park N-S.; Kim M-W.; Langford S.C.; Dickinson J.T. *J. Appl. Phys.* **1996**, 80, 2680-2686.

23. Dove P.M.; Hochella M.F. *Geochim. Cosmochim. Acta* **1993**, 57, 705-714.

24. Gratz A.J.; Hillner P.E.; Hansma P.K. *Geochim. Cosmochim. Acta* **1993**, 57, 491-495

25. Hillner P.E.; Manne S.; Hansma P.K. *Faraday Discuss.* **1993**, 95, 191-197.

26. Liang Y.; Lea A.S.; Baer D.R.; Engelhard M.H. *Surf. Sci.* **1996**, 351, 172-182.

27. Liang Y.; Baer D.R.; McCoy J.M.; Amonette J.E.; LaFemina J.P. *Geochim. Cosmochim. Acta* **1996**, 60, 4883-4887.

28. Liang Y.; Baer D.R. *Surf. Sci.* **1997**, 373, 275-287.

29. McCoy J.M.; LaFemina J.P. *Surf. Sci.* **1997**, 373, 288-299.

30. Pina C.M.; Bosbach D.; Prieto M.; Putnis A. *J. Crystal Growth* **1998**, 187, 119-125.

31. Fubini B.; Bolis V.; Bailes M.; Stone F. *Solid State Ionics* **1989**, 32, 258-272.

32. Deer W.A.; Howie R.A.; Zussman J. *Introduction to the Rock Forming Minerals*, Longman, Harlow, UK, 1992.

33. Blanchard D.L.; Baer D.R.; *Surf. Sci.* **1992**, 276, 27-39.

34. Kenway P.R.; Oliver P.M.; Parker S.C.; Sayle D.C.; Sayle T.X.T.; Titiloye J.O. *Mol. Simul.* **1992**, 9, 83-98.

35. MacInnis I.N.; Brantley S.L. *Geochim. Cosmochim. Acta* **1992**, 56, 1113-1126.

36. Parker S.C.; Kelsey E.T.; Oliver P.M.; Titiloye J.O. *Faraday Discuss.* **1993**, 95, 75-84.

37. Didymus J.M.; Oliver P.; Mann S.; De Vries A.L.; Hauschka P.V.; Westbroek P. *J. Chem. Soc. Faraday Trans.* **1993**, 89, 2891-2900.

38. Ohnesorge F.; Binnig G. *Science* **1993**, 260, 1451-1456.

39. Gilman J.J. *J. Appl. Phys.* **1960**, *31*, 2208-2218.

40. Davey R.J.; Black G.N.; Bromley L.A.; Cottier D.; Dobbs B.; Rout J.E. *Nature* **1991**, *353*, 549-550.

41. Putnis A.; Junta-Rosso J.L.; Hochella M.F. *Geochim. Cosmochim. Acta* **1995**, *59*, 4623-4632.

42. Prieto M.; Fernandez-Gonzalez A.; Putnis A.; Fernandez-Diaz L. *Geochim. Cosmochim. Acta* **1997**, *61*, 3383-3397.

43. Eggleston C.M.; Hug S.; Stumm W.; Sulzberger B.; Dos Santos Afonso M. *Geochim. Cosmochim. Acta* **1998**, *62*, 585-593..

44. Nygren M.A.; Gay D.H.; Catlow C.R.A.; Wilson M.P.; Rohl A.L. *J. Chem. Soc. Faraday Trans.* **1998**, *94*, 3685-3693.

45. Compton R.G.; Brown C.A. *J. Colloid Interface Sci.*, **1994**, *165*, 445.

46. Jordan G.; Rammensee W. *Geochim. Cosmochim Acta* **1998**, *62*, 941-947.

47. T.R. Forester, and W. Smith, *DL_POLY user manual*, (CCLRC, Daresbury Laboratory, Daresbury, Warrington, UK, 1995)

48. Parker S.C.; Price G.D. *Advances in Solid State Chemistry*, **1989**, 1, 295-327

49. P. Dauber-Osguthorpe; D.J. Osguthorpe; P.S. Stern; J. Moult, *J. Comp. Phys.*, **1999**, *151*, 169-189.

Chapter 9

Molecular Statics Calculations of Acid-Base Reactions on Magnetite(001)

James R. Rustad and Andrew R. Felmy

W. R. Wiley Environmental Molecular Sciences Laboratory, Pacific Northwest National Laboratory, Richland, WA 99352

Gas phase proton affinities and acidities are calculated for surface oxygen sites on the hydroxylated magnetite (001) surface. The binding energies of protons are not consistent with the higher acidity observed for the magnetite surface relative to the hematite and goethite surfaces. The pH of zero charge of magnetite, calculated using a linear free energy relationship between the gas-phase and solution pK_as of the iron hydrolysis species, is 9.6. It is therefore proposed that the higher acidity of magnetite is either due to solvent effects or the complex electronic structure of magnetite. We also find that the relaxation of the surface in response to protonation and deprotonation is highly collective, often involving additional acid-base reactions at adjacent sites. These collective effects are as important or more important than, for example, the charge and number of the coordinating metal ions at a given oxide surface site. While this behavior may be represented in a mean field model, we suggest that explicit recognition of these collective interactions may have benefits in aqueous surface chemistry.

The sorption of ions onto oxide surfaces from aqueous solutions is strongly dependent on the pH of the solution. One important measure of the reactivity of oxide surfaces in aqueous environments is the point of zero charge (PZC), defined as the solution pH at which the net charge on the oxide surface is zero in the absence of other specifically absorbing ions. The PZCs of oxide surfaces vary widely with composition. The PZCs of most minerals can be understood, within 1-2 log units, in terms of the size/charge ratios of the constituent metal ions. Oxides with small highly charged cations tend to be more acidic than those with large cations of smaller charge. Thus, Si(IV) ($r_{Si-O}=1.60$ Å) oxides are quite acidic, having PZCs near 2-3; the Ti(IV) ($r_{Ti-O}=1.95$) oxides are somewhat acidic, having PZCs near 5-6; Al and Fe(III) oxides are basic, having PZCs near 9; and MgO has a PZC nearly equal to 12^1.

It has been well established that the pH of zero charge (PZC) of magnetite (Fe_3O_4) is 2-3 pH units below that of the fully oxidized iron oxide materials hematite and goethite[1]. The low value for the PZC of magnetite is unexpected because the addition of some reduced Fe(II) component to the system should, if anything, make the surface more basic. Three hypotheses might explain the low PZC of magnetite: (1) the presence of Fe(II) at the surface allows the Fe(III) sites greater flexibility in accommodating hydroxyl ions, effectively making them more acidic; (2) solvent effects preferentially enhance the acidities of magnetite surface functional groups over those of ferric oxides; (3) the reason for the higher acidity of magnetite is buried in its complex electronic structure and the effect cannot be understood in the context of an ionic model.

This paper focuses on the first of these hypotheses. Molecular statics calculations are carried out on the proton affinities and acidities of individual oxide surface sites on magnetite. The model employed here is an ionic model which was originally designed to calculate structures and energies for hydroxylated/hydrated ferric oxide surfaces. The O-H potential functions were taken from the polarizable, dissociating water model of Halley and co-workers[2], a modified form of the Stillinger-David model[3]. The Fe-O parameters, including the short-range repulsion and charge-dipole cutoff functions, were fit to the Fe^{3+}-H_2O potential surface of Curtiss and co-workers[4]. This model has been used in a variety of applications including ion hydrolysis in solution[5], ferric oxide and oxyhydroxide crystal structures[6], the vacuum termination of hematite (001)[7], monolayers of water on hematite (012)[8,9], and the surface charging behavior of goethite (FeOOH)[10,11] and hematite[9].

More recently, the potentials were applied to magnetite[12]. In that work, it was shown that reasonable structures could be obtained for magnetite and wüstite simply by changing the Fe charge to 2.5+ or 2+ , keeping the same short-range repulsion and charge-dipole cutoff function parameters. Both magnetite and wüstite and Fe(II)-O-OH-H_2O molecules were predicted to have Fe-O bond lengths and lattice constants about 5 percent too large, which is similar to the model errors found for the ferric oxide structures. For example, an Fe^{2+}-O distance of 2.21 Å was predicted for $Fe(H_2O)_6^{2+}$, as compared to the experimental value of 2.15 Å. High-level quantum mechanical calculations on the hexaaquo complex also predict Fe(II)-O bond lengths which are too long, giving 219 pm[13]. Given our good agreement with experiment in the studies referenced above, and the challenges in applying *ab initio* methods to magnetite, it seems reasonable to continue to explore the predictions of this simple model.

Methods

The crystal surface is represented as a 2-D periodic slab of magnetite possessing a center of symmetry. Convergence of the slab energy with respect to thickness was tested in reference (12). We employ a one-layer (12.5 Å) slab in this work. In our calculations we use the average charge of 2.5+ for ions in the octahedral "B" sites and 3+ for the ions in the tetrahedral "A" sites. Above the Verwey transition, distinct Fe^{2+} and Fe^{3+} sites do not exist in bulk magnetite. Spin-sensitive STM studies[14] indicate, however, that distinct Fe^{2+} and Fe^{3+} sites *do* exist at the *vacuum terminated* surface. Here, for the *hydroxylated* surface, we justify using the delocalized model on the grounds that all surface sites retain their full octahedral geometry, and assume that the Verwey transition is similar to that of the bulk.

The proton arrangement on the surface, shown in Figure 1, was previously determined[15]. As shown in the figure, there are 12 surface oxide sites. The sites chosen for addition/removal of protons are denoted by "a" (acid, proton removal) and "b" (base, proton addition). The proton-accepting sites consist of two $Fe^{3+}OH$ (b_4,b_5) sites, one $Fe^{2.5+}OH$ site (b_1), and three $Fe^{2.5+}_2Fe^{3+}O$ (b_2, b_3, b_6) sites. The proton donating sites include three $Fe^{2.5+}_2OH_2$ (a_2, a_5, a_6) sites, two $Fe^{2.5+}_3OH$ sites (a_1, a_4) and one $Fe^{2.5+}_2Fe^{3+}OH$ sites (a_3).

The gas-phase acidities (GPA) and gas-phase proton affinities (GPPA) are calculated by removing two protons (for the GPA) or adding two protons (for the GPPA) to the surface represented in Figure 1. Two protons are involved because the center of symmetry within the slab must be maintained. The energy of the system with the extra protons (or hydroxyls) is minimized and the difference in the final energy is calculated (and divided by 2 because two protons were added/removed). The energy minimizations were carried out with a standard extended Lagrangian molecular dynamics code employing 2-D periodic (x and y) Ewald sums[16] which calculates forces on all the ions as well as "fictitious" forces $-\partial\Phi/\partial\mu$ (where Φ is the potential energy) on the the dipoles. The forces are used in a conjugate gradient energy minimization code attached to the molecular dynamics code. The energy term due to the neutralizing background charge is added in each system for which the ionic charge is not zero.

The potential functions employed are the same as those used in previous papers[5-12]. All ions are assumed to have formal charges: Fe^{3+}, $Fe^{2.5+}$, O^{2-}, and H^+. Oxide ions have inducible point dipoles that respond to every other ion in the system, including the attached protons. The O-Fe and O-H charge-dipole interactions are damped at close range with a function $S(r)$. For the O-O interactions, the following form is used:

$$\phi_{OO} = 1/2\,\Sigma_O\Sigma_{O'} + A_{OO}/r_{OO'}^{12} + B_{OO}/r_{OO'}^{6} + q_Oq_O/r_{OO'} + q_O(\mu_{O'}\cdot r_{OO'})/r_{OO'}^3$$

$$+ 1/2\,\mu_O\,(I - 3r_{OO'}r_{OO'}/r_{OO'}^2)\,\mu_{O'} \tag{1}$$

where $r_{xy} = r_x - r_y$, I is the 3x3 unit matrix, and μ_O is the dipole vector on the oxygen. The functional form of the O-H interaction is:

$$\phi_{OH} = \Sigma_O\,\Sigma_H\,a_{OH}\,exp\,(-b_{OH}\,r_{OH})/r_{OH} + [c_{OH}(r_{OH}-r_{0OH})^2 - d_{OH}(r_{OH}-r_{0OH})]exp(-e_{OH}(r-r_{0OH})^2)$$

$$+ q_Hq_O/r_{OH} + q_H(\mu_O\cdot r_{OH})/r_{OH}^3\,(S(r_{OH})) \tag{2}$$

where

$$S(r_{OH}) = r^3/(r^3 + f(r_{OH})) \tag{3}$$

and

$$f(r_{OH}) = [f_{OH}(r-r_{0OH})exp(-g_{OH}(r-r_{0OH})) + h_{OH}exp(-p_{OH}r_{OH})]/(1+exp(s_{OH}(r-t_{OH})) \tag{4}$$

For water molecules, there is a three-body term of the form

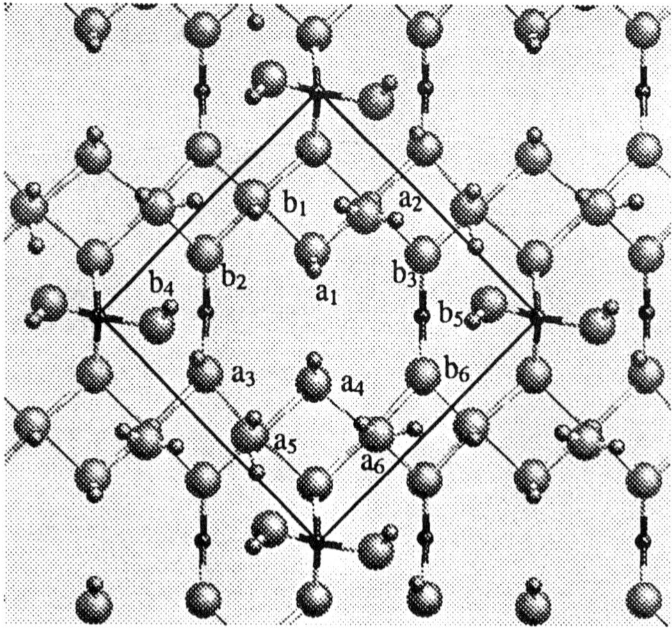

Figure 1. *Proton distribution on the hydroxylated magnetite (001) surface. The unit cell is denoted by the solid line. The surface sites consist of proton donating acid sites a_1-a_6 and proton accepting sites b_1-b_6. Note that the oxygens within the unit cell attached to the tetrahedral irons at the top and bottom vertices of the unit cell are bulk oxygens coordinated by 3 $Fe^{2.5+}$ and 1 Fe^{3+}. As in a_1 and a_4, one of the coordinating $Fe^{2.5+}$ sites lies eclipsed directly beneath the oxygen and is not visible. Also not that $Fe^{2.5+}$ ions lie directly beneath b_1, a_2, a_5, and a_6.*

$$\phi_{HOH} = \Sigma_O \Sigma_H \Sigma_{H'} \, [a_{HOH}(r_{OH}-r_{0OH})(r_{OH'}-r_{0OH})+1/2b_{HOH}(\theta-\theta_0)^2$$

$$+c_{HOH}(r_{OH}+r_{OH'}-2r_{0OH})(\theta-\theta_0)+d_{HOH}(\theta-\theta_0)]exp\{e_{HOH}(r_{OH}-r_{0OH})^2+(r_{OH'}-r_{0OH})^2\} \qquad (5)$$

For the Fe-O interactions the functional form is:

$$\phi_{FeO} = \Sigma_{Fe} \Sigma_O A_{FeO}exp(-B_{FeO}r_{FeO}) - C_{FeO}/r^6_{FeO} + D_{FeO}/r^{12}_{FeO} + q_{Fe}q_O/r_{FeO}$$

$$+ q_{Fe}(\mu_O \cdot \mathbf{r}_{FeO})/r^3_{FeO}(S(r_{FeO})) \qquad (6)$$

where

$$S(r_{FeO})= 1- 1/(exp(E_{FeO}(r-F_{FeO})) \qquad (7)$$

H-H and Fe-Fe interactions are purely coulombic:

$$\phi_{HH} =1/2 \, \Sigma_H \Sigma_{H'} q_H q_H/r_{HH'} \qquad (8)$$

$$\phi_{FeFe} =1/2 \, \Sigma_{Fe} \Sigma_{Fe'} q_{Fe}q_{Fe}/r_{FeFe'} \qquad (9)$$

Oxide ion polarizability contributes $\phi_{pol}=1/2\Sigma_O\mu_O^2/\alpha$ to the potential energy.

Table 1 list the potential parameters in units of e (charge), Å (length) and e^2/Å (energy).

Results

GPPAs and GPAs are reported in Table 2. Upon addition/removal of a proton, it was often found that the relaxation around the defect resulted in proton transfer at other sites. Deprotonation of all the $Fe^{2.5+}OH_2$ sites resulted in surface configurations having the same proton distribution. Therefore all $Fe^{2.5+}OH_2$ have the same acidity. Coincidentally, the acidity of the $Fe^{2.5+}{}_2Fe^{3+}OH$ site is the same as that of the $Fe^{2.5+}OH_2$. The two $Fe^{2.5+}{}_3OH$ sites have somewhat smaller GPA (*larger* proton binding energies). Equilibrium constants were calculated from the GPPAs and GPAs from the linear relationship between calculated deprotonation energies and known log Ks for successive iron hydrolysis as derived in references (10) and (11).

To obtain the PZC from the equilibrium constants in a multisite model we use the expression from reference (10). To derive that expression consider the equilibrium constants K^b_i and K^a_j resulting from the equilibria:

$$S + H^+ \rightarrow SH_i^+; \quad K^b_i=[SH_i^+]/([S][H^+]) \quad (i=1,n) \qquad (10)$$

$$S_j^- + H^+ \rightarrow S; \quad K^a_j=[S]/([S_j^-][H^+]) \quad (j=1,m) \qquad (11)$$

Table I. Parameters for the molecular statics model of the magnetite surface.

A_{OO}	2.02	a_{HOH}	-0.640442
B_{OO}	1.35	b_{HOH}	0.019524
		c_{HOH}	-0.347908
a_{OH}	10.173975	d_{HOH}	-0.021625
b_{OH}	3.69939	e_{HOH}	16.0
c_{OH}	-0.473492	θ_{0HOH}	104.45 deg
d_{OH}	0.088003		
e_{OH}	16.0	A_{FeO}	1827.1435
f_{OH}	1.3856	B_{FeO}	4.925
g_{OH}	0.01	C_{FeO}	-2.136
h_{OH}	48.1699	D_{FeO}	-74.680
p_{OH}	3.79228	E_{FeO}	1.0
s_{OH}	3.0	F_{FeO}	1.8
t_{OH}	5.0		
r_{0OH}	0.9584	α	1.444 $\mathring{A}^3$
		note that the short-ranged	
q_H	1+	FeO parameters are the same	
q_O	2-	for 2.5+ and 3+ sites.	
q_{Fe}	3+ tet. sites		
q_{Fe}	2.5+ oct. sites		

NOTE: When used in conjunction with equations (1-9), energies in $e^2/\mathring{A}$ are generated. For reference, the water molecule at equilibrium geometry has an energy of 3.11595 $e^2/\mathring{A}$; the $Fe(H_2O)^{3+}$ complex has Fe-O = 1.8506 $\mathring{A}$, an induced dipole moment μ_O of 0.232222 $e\mathring{A}$ (positive side toward the Fe), and a binding energy of 0.48598 $e^2/\mathring{A}$ relative to Fe^{3+} and H_2O; the hexaaquo $Fe(H_2O)_6^{3+}$ complex has Fe-O=2.0765 $\mathring{A}$, μ_O=.577682 $e\mathring{A}$ (positive side toward the Fe), and a binding energy of 2.124873 $e^2/\mathring{A}$ relative to Fe^{3+} and 6 H_2O.

Table II. Sites present in Figure 1.

Site	Site Type	Fe-O bond lengths ($\mathring{A}$) on neutral surface	GPA GPPA	LFER pK_a	CD-MUSIC
Acceptor Sites		Bulk: $Fe^{3+}O=1.94$ $Fe^{2.5+}O=2.12$			
b_1	$Fe^{2.5+}OH$	2.09	-340	10.2	11.5
b_2	$Fe^{2.5+}_2Fe^{3+}O$	1.99, 2.06, 1.84	-352	10.5	4. 5
b_3	$Fe^{2.5+}_2Fe^{3+}O$	2.02, 1.97, 1.89	-343	10.3	4. 5
b_4	$Fe^{3+}OH$	1.90	-343	10.3	5.1
b_5	$Fe^{3+}OH$	1.91	-340	10.2	5.1
b_6	$Fe^{2.5+}_2Fe^{3+}O$	2.06, 2.01, 1.85	-352	10.5	4. 5
Donor Sites					
a_1	$Fe^{2.5+}_3OH$	2.13, 2.08, 2.14	307	9.3	10.9
a_2	$Fe^{2.5+}OH_2$	2.17	288	8.7	11.5
a_3	$Fe^{2.5+}_2Fe^{3+}OH$	2.02, 2.21, 2.14	288	8.7	4. 5
a_4	$Fe^{2.5+}_3OH$	2.19, 2.11, 2.10	311	9.4	10.9
a_5	$Fe^{2.5+}OH_2$	2.21	288	8.7	11.5
a_6	$Fe^{2.5+}OH_2$	2.16	288	8.7	11.5

NOTE: Sites are listed by donors and acceptors. The list includes site type as identified by coordinating metal ions and protons, the bond lengths to the coordinating Fe ions, the gas-phase acidity/gas-phase proton affinity (GP/GPPA), the log K for the reaction using the relationship from references (10) and (11), and the prediction of the CD-MUSIC model described in reference (18).

where S represents a neutral surface species, SH_i^+ represents the ith surface state with a proton attached at site i, and S_j^- represents the jth surface state with a proton removed from site j.

From charge balance at the PZC,

$$\Sigma_i[SH_i^+] - \Sigma_j[S_j^-]=0 \tag{12}$$

or

$$\Sigma_i[S][H^+]K_i^b - \Sigma_j[S]/(K_j^a[H^+])=0 \tag{13}$$

or

$$[H^+]^2\Sigma_i\ K_i^b=\Sigma_j 1/K_j^a \tag{15}$$

or

$$pH_{PZC} = 1/2(\log \Sigma_i\ K_i^b\ -\ \log \Sigma_j 1/K_j^a) \tag{16}$$

Applying formula (16), we obtain PZC of 9.6 for magnetite (001), which is similar to that obtained for goethite (α-FeOOH)[10,11] and hematite (α-Fe$_2$O$_3$)[9] (both near 9) in previous work. Therefore, we conclude, in the context of the simple model described above, that the higher acidity of the magnetite surface is not due to enhanced hydroxyl affinities for the Fe^{3+} sites resulting from the greater flexibility of nearby Fe^{2+}-O bonds. Hence, we must look to the other hypotheses concerning the effects of solvation and the complex electronic structure of magnetite, and possible compositional heterogeneities at the magnetite surface not characteristic of bulk magnetite.

Discussion

Currently, the only comprehensive means of evaluating pK_as for multisite oxide surfaces is the MUSIC (MUltiSIte Complexation) model originally proposed by Hiemstra and co-workers[17]. This model was based on correlations between the size/charge/bond order characteristics of solution complexes and their known pK_as. This model was later reparameterized[18] to take into account distance dependence in much the same manner as the bond valence model of I. D. Brown[19], as well as solvent effects. In Table 2 we give the pK_as calculated for the surface sites on magnetite (001) using the CD MUSIC model. In this model we use R_0 (Fe^{3+})=1.759, R_0 ($Fe^{2.5+}$)=1.7465 (the average of 1.759 for Fe^{3+} and 1.734 for $Fe^{2.5+}$), and the experimental bond distances in bulk magnetite of 2.07 Å for $Fe^{2.5+}$-O and 1.87 Å for Fe^{3+}-O. We also assume that, for each species, the site accepts one additional hydrogen bond from solution after deprotonation. It is evident from Table 2 that there are significant differences between our model and the CD-MUSIC model. Certain sites having almost identical gas-phase proton affinities, such as $Fe^{2.5+}OH$ (pK_a: 11.5; GPPA: -340 kcal/mol) and $Fe^{2.5+}_2Fe^{3+}O$ (pK_a: 4.5; GPPA:-343 kcal/mol), have widely different pK_as in the CD-MUSIC model. The same is true of the donor sites $Fe^{2.5+}OH_2$ (pK_a: 11.5; GPA: 288 kcal/mol) and $Fe^{2.5+}_2Fe^{3+}OH$ (pK_a: 4.5; GPA: 288 kcal/mol). We have found repeatedly for the ferric oxides that the gas-phase proton affinities of the triply coordinated Fe_3O sites are larger that would be expected based on the MUSIC model[10]. Another factor contributing to the enhanced acidity of the magnetite surface in the MUSIC model is short tetrahedral Fe-O bond distance. Note that the $Fe^{3+}OH$ sites have the same proton affinity as the $Fe^{2.5+}OH$ sites (GPPA= -340 kcal/mol).

In the particular case of the CD-MUSIC model, the low values for the surface pK_as are dependent on the assumption that each site accepts a hydrogen bond from solution after deprotonation. If it is assumed that, for example, these relatively low-lying sites are *not* accessible to the solvent, and therefore do not accept a hydrogen bond on deprotonation, the pK_a for the $Fe^{2.5+}_2Fe^{3+}OH$ site is raised four log units to 8.5.

Sverjensky and Sahai[20] have published a single-site model which reproduces the PZCs of a wide variety of oxides and silicates. Their model, includes contributions from both the Pauling bond strength per angstrom at the oxide site and from the dielectric polarization within the adsorbing mineral. Their model does indeed predict a larger acidity for magnetite than for goethite or hematite. The reason for this is the very high dielectric constant used for magnetite in their correlations which show increasing acidity with increasing dielectric constant. A physical interpretation of the positive correlation of the mineral dielectric constant and the surface acidity can be understood in the following way: in terms of their effective two-pK model, the canonical surface site SOH changes to SO^- upon donating a proton and to SOH_2^+ upon accepting a proton. The solvation contribution from the "water side" of the interface was assumed to be constant from mineral to mineral, and has no effect on the acidity. The polarization contribution from the "mineral side" of the interface contributes favorably to both proton adsorption and desorption by responding to the positively charged and negatively charged products. However, the negative charge at the SO^- site is located at a closer distance to the mineral and is therefore *more* effective than the SOH_2^+ site in polarizing the surface. Because of the greater stabilization of the SO^- site, the surface becomes more acidic with increasing dielectric constant.

In the particular case of magnetite, the high dielectric constant results from the mobility of the electrons on the octahedral B sites. We have neglected the possible rearrangement of the ferrous/ferric charge distribution in our calculations, and have consistently used the "2.5+ oxidation state" to model the octahedral B ions. It seems likely that charge rearrangement in the B layer ions during acid-base reactions could have a significant effect on the acidities of the surface oxide sites. Calculating this effect would require sophisticated electronic structural models of magnetite. One requirement of such a model would be a correct low temperature (T < T_{Verwey}) ordering of Fe^{3+} and Fe^{2+} sites[21]. Assessment of the intuitive, if simplistic, physical basis suggested above for the model of Sverjensky and Sahai[20] must, in the case of magnetite, await the successful modelling of the low temperature structure.

As mentioned above, our calculations have indicated the collective nature of the close coupling between acid-base reactions on adjacent sites. The fact that acid-base reactions are induced at nearby sites by deprotonation/protonation reactions suggests a subtle but important change in the molecular interpretation of the 2 pK model with the three canonical states SO^-, SOH, and SOH_2^+. These sites are perhaps best interpreted not as protonation states of individual metals S, but as protonation states of larger regions, such as unit cells, on the surface. One pivotal message of this work is that protonation/deprotonation energetics are influenced as much by proton rearrangement at adjacent sites as by the charge/size ratio of the metal ion (witness the equality in the proton affinities of the $Fe^{3+}OH$ and $Fe^{2.5+}OH$ sites in Table 2). This type of approach may also have benefits in terms of accounting for the entropy changes attendant upon surface acid-base reactions as reflected, for example, in the number of states available for positively charged, negatively charged, and neutral surface unit cells. The qualitative importance of the collective proton rearrangements of surface acid base reactions are unlikely to change as improvements to the model are made to account for the complex electronic structure of magnetite, and will probably become even more significant as interfacial solvent effects are taken into account.

Acknowledgement

This work was supported by the U.S. Department of Energy, Office of Basic Energy Sciences, Engineering and Geosciences Division, contract 18328. Pacific Northwest Laboratory is operated for the U.S. Department of Energy by Battelle Memorial Institute under Contract DE-AC06-76RL0 1830. We are grateful to the National Energy Research Supercomputing Center for a generous grant of computer time.

References

1) Lykelma, J. *Fundamentals of Interface and Colloid Science, Vol. II: Solid-Liquid Interfaces*, Academic Press, London, 1995.

2) Halley, J. W.; Rustad, J. R.; Rahman, A. *J. Chem. Phys.* 1993, *98*, 4110.

3) Stillinger, F. H.; David, C. W. *J. Chem. Phys.* 1978, *69*, 1473.

4) Curtiss, L. A.; Halley, J. W.; Hautman, J.; Rahman, A. *J. Chem. Phys.* 1987, *86*, 2319.

5) Rustad, J. R.; Hay, B. P. ; Halley, J. W. *J. Chem. Phys.* 1995, *102*, 427.

6) Rustad, J. R.; Felmy, A. R.; Hay, B. P. *Geochim. et Cosmochim. Acta* 1996, *60*, 1553.

7) Wasserman, E.; Rustad, J. R.; Hay, B. P.; Halley, J. W. *Surface Science* 1997, *385*, 217.

8) Henderson, M. A.; Joyce, S. A.; Rustad, J. R. *Surface Science* 1998, *417*, 66.

9) Rustad, J. R.; Wasserman, E.; Felmy, A. R. *Surface Science* 1999, *424*, 28.

10) Rustad, J. R.; Felmy, A. R.; Hay, B. P. *Geochim. et Cosmochim. Acta* 1996, *60*, 1563.

11) Felmy, A. R.; Rustad, J. R. *Geochim. et Cosmochim. Acta* 1998, *62*, 25.

12) Rustad, J. R.; Wasserman, E.; Felmy A. R. *Surface Science Letters* 1999, *432*, L583.

13) Akesson, R.; Pettersson, L. G. M.; Sandstrom, M.; Wahlgren, U. *J. Am. Chem. Soc.* 1994, *116*, 8691.
14) Wiesendanger, R.; Shvets, I. V.; Buergler, D.; Tarrach, G.; Guentherodt, H. J.; Coey, J. M. D.; Graeser, S. *Science* 1992, *255*, 583.
15) Rustad, J. R.; Wasserman, E.; Felmy, A. R. *in* Meike, A., Gonis, A.; Turchi, P. E. A. and Rajan, K.; Ed.; *Properties of Complex Inorganic Solids, Volume II* Kluwer Academic/Plenum Publishers, Dordrecht, The Netherlands
16) Parry, D. E. *Surface Science* 1976, *54*, 195.
17) Hiemstra, T.; VanRiemsdijk, W. H.; Bolt, G. H. *J. Colloid Interface Sci.* 1989, *133*, 91.
18) Hiemstra, T.; Venema, P.; VanRiemsdijk, W. H. *J. Colloid Interface Sci.* 1996, *184*, 680.
19) Brown, I. D.; Altermatt, D. *Acta Cryst. B* 1985, *B41*, 244.
20) Sverjensky, D.; Sahai, N. *Geochim. et Cosmochim. Acta* 1996, *60*, 119.
21) Iida, S.; Mizushima, M.; Mizoguchi, K.; Kose, K.; Kato, K.; Yanai, K.; Goto, N.; Yumoto, S. *J. Appl. Phys.* 1982, *53*, 2164.

Chapter 10

Electronic Structure and Chemical Reactivity of Metal Oxides–Water Interfaces

Thanh N. Truong[1], Michael A. Johnson, and Eugene V. Stefanovich

Henry Eyring Center for Theoretical Chemistry, Department of Chemistry, University of Utah, 315 South 1400 East, Room Dock, Salt Lake City, UT 84112
[1]Corresponding author: Truong@chemistry.utah.edu

We present our recent progress in studying electronic structure and chemical reactivity of metal oxide-water interface. Particularly, we examined both molecular adsorption and dissociative chemisorption of water at the MgO(100)-water interface using the CECILIA (Combined Embedded Cluster at the Interface with LIquid Approach) model. This model combines advances in dielectric continuum solvation models for describing polarization of the liquid with the embedded cluster approach for treating interactions in the solid. Comparisons with results from detailed molecular dynamics and Monte Carlo simulations and with experiments were made.

Introduction

Ion sorption and chemical reactions at interfaces between metal oxides and water are central features in many natural and industrial processes. Examples include transportation of groundwater contaminants, electrode phenomena, corrosion, and dissolution. For geochemistry and atmospheric chemistry, surfaces of metal oxides are of particular interest as these compounds are major components of rocks, soils, and airborne dust particles. For many oxides it has been found that water molecules dissociate upon contact with the surface, forming various types of surface hydroxyl groups. It is also well established that these hydroxyl groups play a decisive role in many chemical properties of oxide surfaces, including ion sorption, dissolution, and catalytic activity [1].

MgO surfaces and their interactions with water provide an excellent example for the discussion of this chapter due to its extensive theoretical and experimental literatures. However, the primary reason for choosing the MgO-water system is the previously unsettled problem regarding chemical reactivity of the most stable MgO(100) surface toward water molecules that the recent theoretical method presented in this chapter had help to resolve. In particular, many previous experimental [2-5] and theoretical [6-10] studies have found that water molecules do not dissociate upon adsorption on the MgO(100) crystal surface from vacuum. However, there are several experimental indications that water can dissociate at the MgO(100)-water interface. For example, the naturally occurring transformation of mineral periclase (MgO) to thermodynamically favored brucite ($Mg(OH)_2$) implies surface hydroxylation as an intermediate step. The commonly used argument to explain this transformation requires the involvement of low-coordinated surface sites [3, 11-14] or high Miller index surface planes [9, 15-17]. It is worthwhile to note that such defective structures are minority sites at the MgO surface. Moreover, this argument does not explain the independence of the initial rate of dissolution on the presence of defects and the formation of (100) facets upon immersion of MgO crystallites in water [18]. In addition, experimental studies show that the (100) surface dissolves in liquid water [19], and the rate of dissolution increases with increasing acidity [20]. It is also known that the presence of water can radically alter reaction mechanisms and kinetics at MgO surfaces [21]. All these facts suggest that hydroxylation of the MgO(100)-water interface is quite likely.

In this chapter, we discuss applications of our recently proposed quantum embedded cluster methodology, called CECILIA (Combined Embedded Cluster at the Interface with LIquid Approach) [22, 23] to address the electronic structure and chemical reactivity at the MgO-water interface [24, 25]. The CECILIA model combines advantages of the embedded cluster method [26-28] with the dielectric continuum method for solvation to provide an accurate description of bond-forming and -breaking processes and interactions of adsorbates and surface defects with the crystal lattice and solvent. Since the details of the CECILIA model have already been discussed in several previous reports [22, 23], we briefly discuss it below.

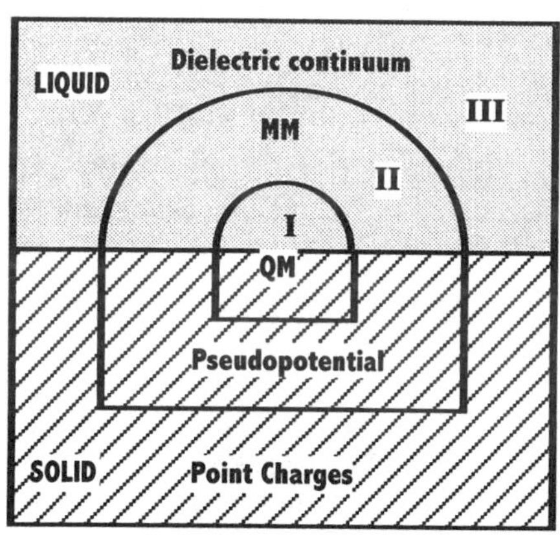

Figure 1: Schematic of the CECILIA model

A Physical Model of the Solid-Liquid Interface

In the CECILIA approach, the whole system (surface defect + crystal + solvent) is divided into three main regions (Figure 1) designed to maximize the chemical accuracy while keeping the problem tractable by modern computers.

I. The innermost QM (quantum mechanics) region, where chemistry occurs, is treated by an accurate *ab initio* molecular orbital (MO) or density functional theory (DFT). Normally, the QM cluster may consist of several lattice atoms near the defect site, the adsorbate and a couple of water molecules making strong hydrogen bonds with the surface complex. The size of this region, level of theory, basis set, and use of effective core pseudopotentials are dictated by the specific problem and available computational resources.

II. The buffer or MM region normally includes several dozen atoms in the crystal lattice surrounding the QM cluster and several solvent molecules. This region is designed to describe short-range forces between nuclei and electrons in the QM cluster and surrounding medium.

III. The peripheral zone containing point charges and a dielectric continuum ensures correct Madelung and long-range solvent polarization potentials in the quantum cluster region. This is important for an accurate representation of the cluster electron density, correct positions of the crystal electronic band edges with respect to vacuum, and redox potentials of molecular solutes. As described in several of our previous studies [23, 29, 30] several hundred point charges are often sufficient to reproduce the Madelung potential in the cluster region with an error of less than 1%. This is done by dividing the Madelung potential into two components. The component from region that is closed to the buffer zone is represented by a set of point charges located at the lattice positions. The other component is from the remaining extended crystal. This can be represented by a set of surface charges using our SCREEP (Surface Charge Representation of External Electrostatic Potential) method proposed earlier [31]. A self-consistent treatment of the solvent polarization can be achieved by using dielectric continuum solvation methods [32, 33]. We adopted the GCOSMO method documented in the literature [34-37]. It is important to point out that the COSMO boundary condition [38] does not strictly require the boundary to be a closed surface. Dispersion-repulsion contributions to the solvation free energy were calculated using Floris and Tomasi's method [39], in conjunction with OPLS force field parameters [40]. Cavitation energy was calculated using a method suggested by Pierotti [41], Huron and Claverie [42].

Computational Details

We used the cluster shown in Figure 2 to model the MgO(001) surface. For computational feasibility, ionic cores were approximated by effective core pseudopotentials (ECP) [43]. We used the standard valence CEP-31++G(d,p) basis set on the atoms of the water molecule. The CEP-31G(d) basis set was used on the

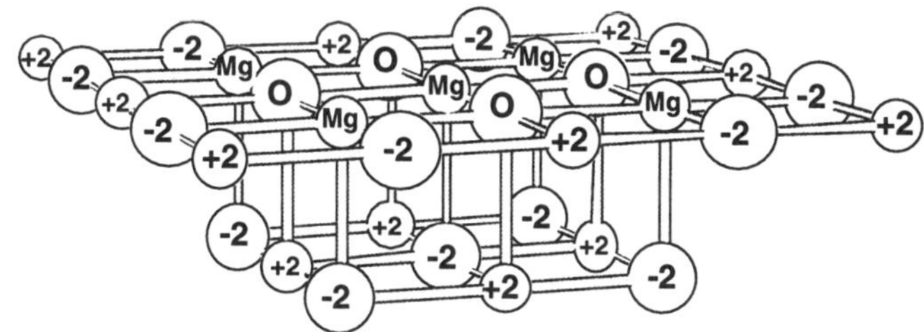

Figure 2: Cluster model of the MgO(100) surface. This cluster is embedded in the field generated by 222 additional point changes (not shown).

four oxygen ions (labeled "O") nearest to the central Mg ion. The additional diffuse functions on the water oxygen atom allow for more accurate description of adsorbate-surface interaction and were found to have only a small effect on the density of states spectra even for the dissociated chcmisorption case. Oxygen ions marked as "-2" in Figure 2 were modeled as point charges (q_0 = -2) without basis functions. The CEP-31G basis set was placed on the five Mg ions at the surface (labeled "Mg"); other Mg ions in the cluster (labeled "+2") were approximated by bare pseudopotentials without basis sets. In order to represent the rest of the crystal, the cluster described above was embedded in the field generated by 222 lattice point charges of ±2 (not shown in Figure 1) so that the entire system (cluster + point charges) consisted of four stacked 8 x 8 layers resulting in an 8 x 8 x 4 slab. This finite lattice has been shown to provide an accurate Madelung potential at the (001) rock salt crystal surface without the need of using SCREEP surface charges [29, 30].

We used a Solvent-Excluding Surface [44], and the cavity boundary generated by the GEPOL93 algorithm [44] was truncated so that only adsorbed atoms and surface ions of Figure 2 were solvated. Atomic radii for cavity construction were taken from our previous work: 1.172 Å for H, 1.576 Å for O [45]. The atomic radius for Mg (1.431 Å) was fitted to the experimental hydration free energy of the Mg^{2+} ion.

Geometry optimizations, where adsorbate atoms were fully relaxed and all surface ions were held fixed at ideal lattice positions, were performed using the Hartree-Fock (HF) theory. Electron correlation corrections were estimated by performing single-point second order Møller-Plesset perturbation theory (MP2) calculations at HF optimized geometries.

In characterizing the surface electronic structure of cluster and adsorbed water configurations considered in this work, we present results for the electronic density of states (DOS). This information can be useful for qualitative analysis of data collected in electron spectroscopy experiments [13, 46-49]. As suggested earlier [47], to attain the best agreement between calculated density of states and experimental UPS and MIES spectra for MgO, our DOS graphs were generated by smoothing of orbital energy levels with Gaussian functions having a width of 1.0 eV at half-maximum.

All calculations were performed using our locally modified GAUSSIAN92/DFT computer code[50].

Results and Discussion

We focus the discussion below only on the molecular and dissociate adsorption of water at the MgO(100)-water interface. Adsorption at the MgO(100)-vacuum interface has been addressed in numerous reports and thus does not need to be included here. Of particular interest to the discussion below are the comprehensive MD and MC simulation results on the structure and dynamics of molecular water at the MgO-water interface performed by McCarthy et al.[8] and the experimental metastable impact electron spectroscopy (MIES) results for the detection of hydroxyl species from dissociative chemisorption of water at the water covered MgO(100) surface done by Goodman and co-workers [51]. In these simulations, 2-D periodic boundary condition with unit cells of 64 and 128 water molecules was used. This corresponds to several layers of water above the MgO surface. Interactions between water molecules were represented by the SPC potential force field whereas interactions between water and MgO surface were described by a force field fitted to results from *ab initio* periodic electronic structure calculations. In MIES, metastable excited helium atoms were utilized to eject electrons from the substrate surface. In the case of insulating surfaces, the intensity of the ejected electrons versus their kinetic energy gives a direct image of the density of occupied states of the surface. Since the metastable helium atoms approach the surface with thermal kinetic energy of about 200 meV this technique is nondestructive and highly surface sensitive, thus it is able to distinguish molecular and dissociative adsorption species of water at the water-covered MgO(100) interface.

Molecular adsorption

The average structure of H_2O adsorbed at the MgO-water interface (see Table 1) is slightly different from that at the MgO-vacuum interface. The noticeable difference is an increase in the tilt angle by 3.3 degrees. This is in reasonable agreement with calculations by McCarthy et al.[8] who found that in going from the MgO-vacuum to the MgO-water interface, the tilt angle changes from 105 to 107 degrees. In this study, we were not able to reproduce another configuration corresponding to the tilt angle of 60 degrees as observed in the angular distribution of near-surface water molecules [8]. This orientation probably appears as a result of cooperative interactions of several water molecules that were not included in our model. In a more recent study [52] using a 3D extended Reference Site Interaction Model (RISM) model, we were able to predict the later configuration in agreement with the MC simulations. We also found that the liquid water modeled as a dielectric continuum has very little effects on the electronic structure of the MgO surface if only molecular adsorption is assumed.

To have a better understanding on the balance of different interactions at the MgO-water interface, we plot the potential of mean force for moving the adsorbed water at the interface to the bulk liquid as a function the height of the water oxygen above the surface (O_Z). It is interesting to note that the adsorbed H_2O at the interface corresponds to a local minimum on the free energy surface but has 1.8 kcal/mol higher

Table 1: Structure (Å, degrees) and MP2 Adsorption Energy (kcal/mol) for a Water Molecule at the MgO(001) Vacuum and Aqueous Interface.

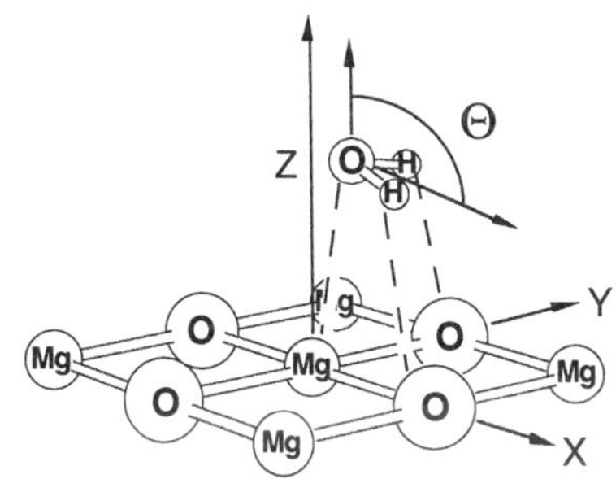

Parameter[a]	Vacuum Interface	Aqueous Interface
O-H Bond Length	0.954	0.956
H-O-H Angle	105.35	105.30
O_z	2.32 (2.02)[b]	2.31 (2-3)[b]
$O_x = O_y$	0.30 (0.54)[b]	0.32
Tilt Angle θ	106.7 (105)[b]	110.0 (107 or 60)[b]
Binding Energy	+14.2 (+17.3)[b]	-1.8

[a]See figure above

[b]Calculated in ref. [8].

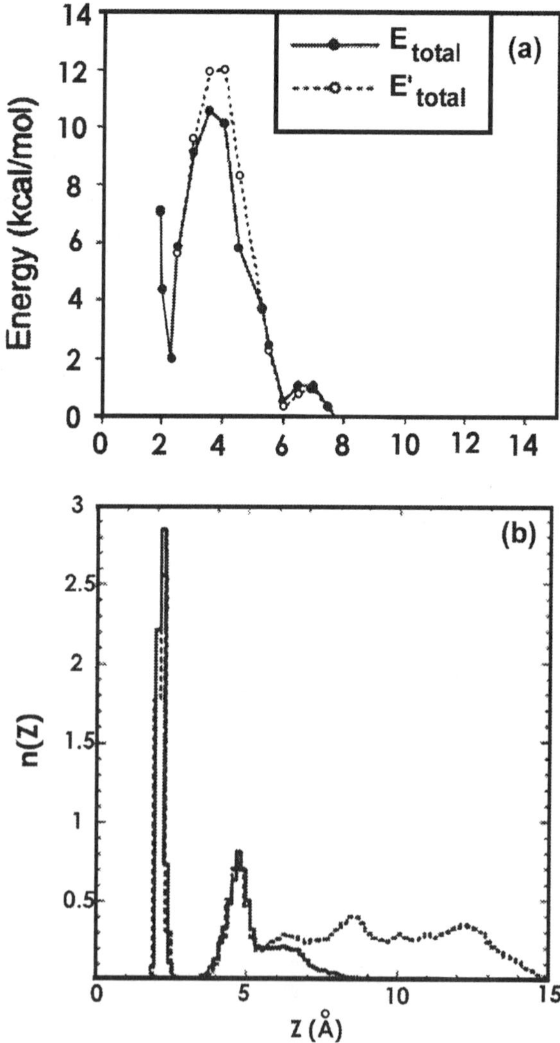

Figure 3: Characteristics of the MgO(001)-water interface as functions of distance (z) between water oxygens and the surface plane. (a) Energy profile $(E_{total}(O_z))$ *for a water molecule adsorbed at the interface calculated using the CECILIA model. (b) Density profiles previously obtained from molecular dynamics simulations (Adapted with permission from ref. [24])*

in energy than that for H_2O in bulk liquid and a barrier of about 9 kcal/mol to move the bulk as shown in Figure 3. This provides an explanation for why adsorbed water molecules rarely exchange with the bulk the MD simulations. However, the origin for such an observation appears to be much more complicated than the simple energetic argument given by McCarthy and coworkers that water molecules confined to the interface are due to the strong water-surface attraction.

We have performed a detailed analysis of the various contributions to the potential of mean force shown Figure 3 by decomposing the interfacial adsorption system in terms of three mutually interacting systems: water molecule, crystal surface, and liquid. Assuming that the total interaction energy is additive, we can write

$$E'_{\text{total}}(O_z) = E_{H_2O-\text{surface}}(O_z) + E_{H_2O-\text{liquid}}(O_z) + E_{\text{surface}-\text{liquid}}(O_z). \quad (1)$$

The three individual contributions to $E'_{\text{total}}(O_z)$ can be estimated from three separate calculations. In each of these calculations we fixed the geometry of the water molecule to be the same as in the case of $E_{\text{total}}(O_z)$. The three interaction curves, as well as the sum represented by $E'_{\text{total}}(O_z)$ above, are shown in Figure 4.

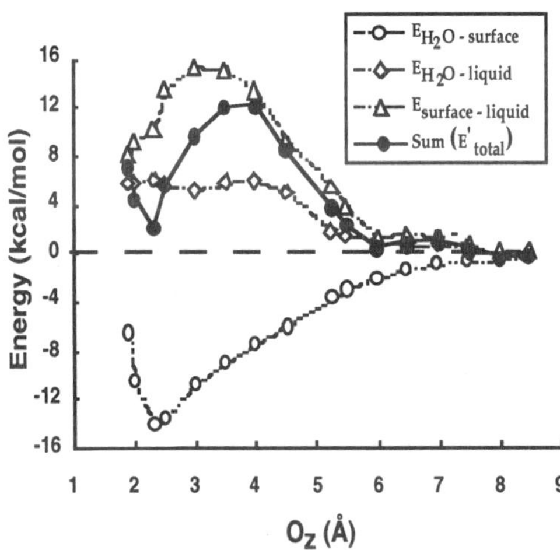

Figure 4: The total interaction energy (solid line) of a water molecule at the MgO-water interface as the sum of three distinct interactions (see eq 1). (Adapted with permission from Ref. [24])

For $E_{H_2O-\text{surface}}(O_z)$ we calculated the energy profile for molecular water desorption from the bare MgO surface represented by the embedded cluster without the dielectric continuum present. This gives a familiar interaction curve with a deep minimum of -14 kcal/mol.

132

For $E_{H_2O-liquid}(O_z)$, the crystal lattice was removed while the dielectric cavity surrounding the surface atoms (at a distance of 1.4-1.6 Å from the crystal surface) and the water molecule remained. This type of calculation resembles the situation that exists when a water molecule penetrates into bulk liquid through the liquid-gas interface. Energy gradually decreases as H_2O penetrates deeper into the bulk liquid and the difference is the free energy of solvation of water.

The curve for $E_{surface-liquid}(O_z)$ was obtained in calculations where the water molecule was absent but the shape of the dielectric cavity above the MgO surface was the same as if the water molecule were present. This plot represents changes in hydration energy of the MgO surface cluster as the shape of the hydration cavity varies in the process of H_2O desorption. $E_{surface-liquid}(O_z)$ has a broad maximum extending from 2 to 6 Å outward from the crystal surface. This barrier originates from a reduction in electrostatic hydration of the MgO surface and a larger cavitation energy due to the increased size of the dielectric cavity when a water molecule moves away from the surface.

As seen in Figure 4 the sum of the three energy components (shown as a broken line) indeed serves as a good approximation to the total interaction energy at the MgO-water interface. At small values of O_z ($\leq$ 2 Å), the leading contribution comes from the repulsive part of the $E_{H_2O-surface}(O_z)$ interaction. $E_{surface-liquid}(O_z)$ is largely responsible for the energy barrier on the $E_{total}(O_z)$ curve. Existence of this barrier prevents water molecules in the first adsorbed layer from exchanging freely with bulk water. In addition, instead of the strong water-surface attraction at the interface we found that due to the combined repulsive effect of $E_{H_2O-liquid}(O_z)$ and $E_{surface-liquid}(O_z)$, water molecules are thermodynamically preferred to be in the bulk.

From the analysis above it follows that energy profiles for other lateral positions of adsorbed water (as well as the energy profile averaged over O_x and O_y) will be qualitatively similar to that presented in Figure 3(a). This allows us to conclude that the average density of water molecules near the interface should anti-correlate with the energy plot in Figure 3(a), i.e., minima on the energy plot correspond to maximum density of particles, and maxima on the energy plot indicate regions more likely to be void of water molecules. Such behavior is clearly seen from comparison of Figure 3(a) with the plot in Figure 3(b), where we have reproduced the density profile for water molecules near the MgO surface obtained in molecular dynamics simulations. The local energy minimum at 2.3 Å corresponds to a narrow sharp density peak at oxygen-surface distances of 2-2.5 Å. The energy barrier at 3-4 Å corresponds to the gap in the density profile at O_z = 3-4 Å. Both profiles of Figure 3 indicate a homogeneous phase of water exists beyond 7 Å of the MgO(001) surface. Although there is some quantitative discrepancy, the overall qualitative agreement between the free energy profile from CECILIA calculations and the water density profile from molecular dynamics simulations is remarkable.

A. Dissociative Adsorption

The key advantage of the CECILIA model is in its ability to study reactive processes at the solid-liquid interface. Using the same procedure as for studying molecular adsorption, heterolytic dissociation of water into charged OH- and H+ at the MgO(100)-water interface is predicted to be energetically favorable.   In fact we found a global minimum corresponding to the formation of two hydroxyl groups as shown in Figure 5.   Here, O_WH^- is bound to the lattice Mg^{2+} ion, and the proton makes a strong bond with the lattice oxygen ion whereby its bond with oxygen from the water molecule is virtually broken (O-H distance is 2.21 Å). However, some residual attraction between OH^- and H^+ species remains which keeps the HOH angle at about 98.9 degrees, close to the value found for the free water molecule.   This minimum has the adsorption energy of 11.0 kcal/mol, that is 21.0 kcal/mol lower

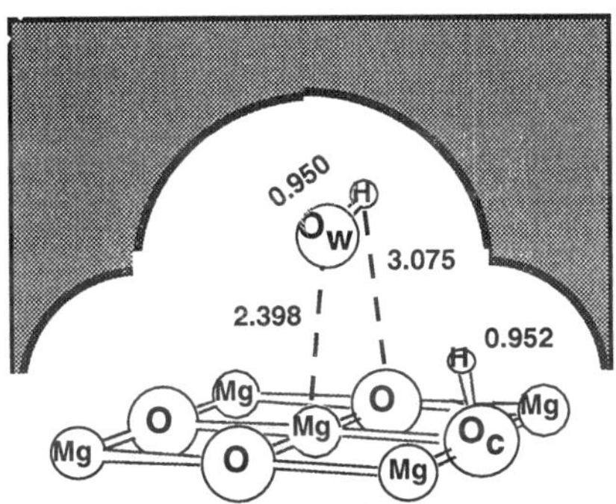

Figure 5: Optimized geometry for dissociated water at the MgO(100)-water interface ([MgO]H⁺OH⁻). Bond lengths are given in Å.

than for a separated adsorbed OH^--H^+ pair. Thus our calculations predict that OH^- and H^+ can exist at the MgO-water interface as nearest neighbor species, and their diffusion away from each other is not likely.

Figure 6 presents the calculated densities of states for some adsorbate configurations. To facilitate comparison with the MIES experiment, the reference zero of energy is also at the Fermi level of the molybdenum substrate. In Table 2 we list the positions of the DOS features with reference to the top of the valence band.

Theoretical positions were determined by defining the highest eigenvalue in the O(2p) band as the valence band edge. Energy differences were then taken between this reference point and the levels of interest. Shown in parentheses in Table 2 are the experimental positions of DOS features relative to the O(2p) band edge.

Table 2: Positions of DOS Features (eV) with Respect to the Top of the Valence Band. Experimental Numbers are in Parentheses.

Feature	Fig. 6b $O_cH^-(g)$	6c $O_cH^-(aq)$	6d $O_wH^-(g)$	6e $O_wH^-(aq)$	6f $O_wH^-(aq)$	6g H_2O $O_cH^-(aq)$
$1b_1$						4.3 (3.4)
$3a_1$						5.8 (5.4)
$1b_2$						9.6 (9.8)
$1\pi_w$			-2.2	2.4	1.5 (1.6)	
$3\sigma_w$			2.2	5.5	5.3 (6.1)	
$1\pi_c$	6.8	5.5			6.2	
$3\sigma_c$	10.0	8.6			9.5	

Figure 6a shows the DOS for the clean MgO(100) surface and provides a base-line for comparison to other adsorption complexes. Upon adsoprtion of H^+ (see Figure 6b), the entire spectrum shifts to higher electron binding energies relative to the MgO(100) surface. As was explained in previous studies [15, 53], this shift is due to the positive electrostatic potential generated in the cluster by the presence of a proton. The local electronic structure of the proton adsorbed on the crystal oxygen (O_c) is similar to that in an OH^- ion. This adsorption complex gives a $2\sigma_c$ peak below the oxygen 2s valence band and a double-peak structure below the O(2p) valence band [15, 54]. The latter peaks are labeled as $3\sigma_c$ and $1\pi_c$ according to the molecular orbital classification in OH^-. In the presence of water, polarization induces a negative potential in the vicinity of the H^+ adsorption site. This reduces the splitting of H^+ induced features from the crystal bands (see Figure 6c).

In agreement with other calculations [15, 54], adsorption of OH^- produces a structure in the band gap region near the top of the O(2p) valence band (labeled $1\pi_w$ in Figure 6d). The $2\sigma_w$ and $3\sigma_w$ features are hidden inside the valence bands. In the presence of water (Figure 6e), polarization induces a positive potential around OH^-, so adsorption-induced levels shift to higher binding energies and $2\sigma_w$ and $3\sigma_w$ levels appear below the oxygen 2s and 2p valence bands, respectively. In this case the $1\pi_w$ peak is inside the oxygen 2p valence band.

The DOS for water dissociated at the MgO-water interface (Figure 6f) exhibits features of both adsorbed H^+ and OH^-. All peaks, except for $1\pi_w$, are well separated from crystal bands, and the $3\sigma_w$ and $1\pi_c$ peaks overlap. In an attempt to understand how much the dielectric solvent contributes to the DOS of dissociated water, we performed a single-point calculation, at the geometry shown in Figure 5, with the solvent removed. Note that this structure does not correspond to a stable minimum at the solid-vacuum interface. The resulting DOS was virtually the same as with the solvent present (Figure 6f). This result indicates that water solvent is important for stabilizing the hydroxyl pair but does not significantly change the electronic structure of dissociated water at the MgO(100)-water interface. Recall that similar results were found for the molecular adsorption case.

The calculated DOS helped to interpret the new features on the MIES spectra upon raising the temperature on multilayer water covered MgO(100) thin film beside those ($1b_2$, $3a_1$ and $1b_1$) from the molecular adsorbed waters as shown in Figure 7. These features correspond to hydroxyl groups on the surface. We can determine from the calculated DOS the approximate nature of the hydroxyl groups that exist on this MgO surface. There are essentially three types of hydroxyls to consider: (i) a protonated surface oxygen ion; (ii) an adsorbed hydroxyl group; (iii) both (i) and (ii) in close proximity and in the presence of a solvent as in Figure 5. The calculated DOS of a type (i) hydroxyl group (Figure 6b and 6c) is distinguished by a doublet that is significantly displaced to higher binding energies from the oxygen 2p band. This is not consistent with the new features of the MIES spectra. A type (ii) hydroxyl group at the MgO-vacuum interface is characterized by a level in the band gap (Figure 6d), and the MIES spectra show no such levels. However, the positions of the hydroxyl peaks ($3\sigma_w$ and $1\pi_w$) in a type (iii) hydroxyl complex (Table 2 and

136

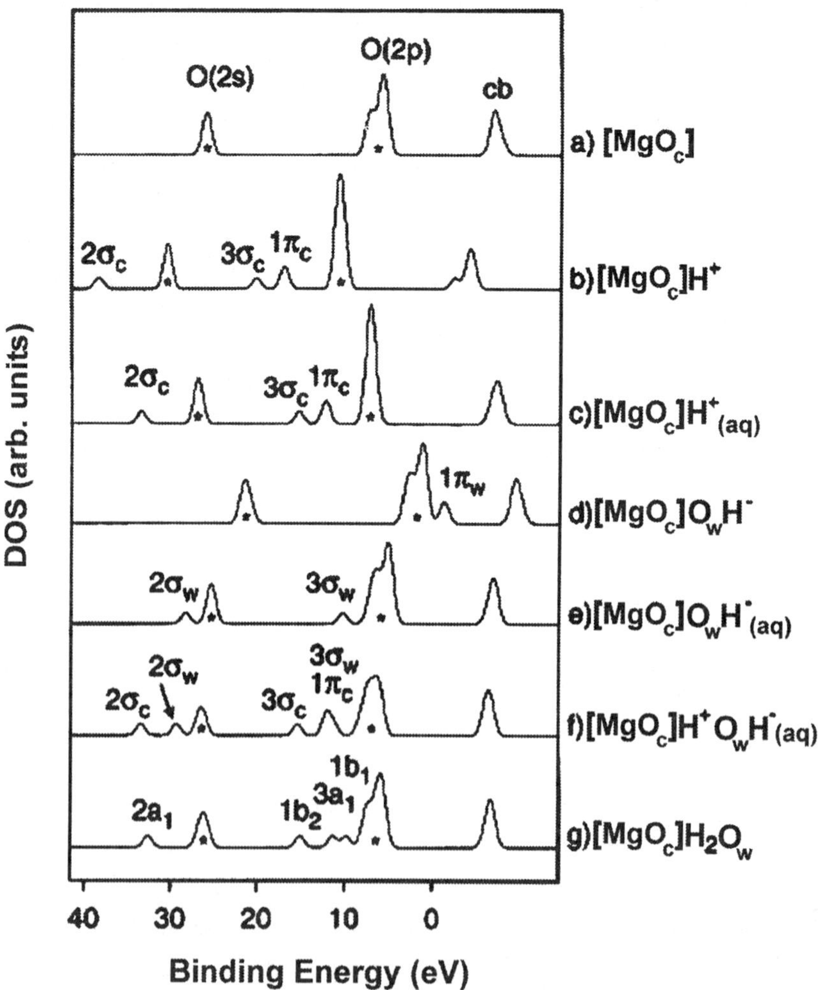

Figure 6: Calculated density of states. The conduction band states are labeled "cb", and an asterisk is used to mark the position of O(2s) and O(2p) crystal bands. "c" and "w" notations indicate features due to the O_cH^- and O_wH^- hydroxyls, respectively. (Adapted with permission from Ref. [51])

Figure 6f) are in good agreement with the MIES spectra of Figure 7. Therefore, we have assigned the two new features of the MIES spectra to the $3\sigma_W$ and $1\pi_W$ hydroxyl levels.

Molecular dynamics simulations indicate the absence of significant structure in the liquid phase beyond the first two layers of water adsorbed on MgO [8]. Therefore, water coverage of 3 ML, as in the MIES experiments, is believed to be sufficient to create a condition near the MgO crystal surface that is very similar to the MgO-water interface environment. In this multilayer coverage regime, our results indicate that water molecules dissociate at the interface; however, the hydroxyl products of such a dissociation are not seen in our spectra due to the surface sensitivity of MIES. Rather, water molecules at the top of the multilayer are visible. Upon raising the temperature, molecular water desorbs and hydroxyls become exposed at about 155 K.

However, it is difficult to say how many water molecules around the OH^--H^+ pair are sufficient to keep this configuration stable. In principle, a layer of adsorbed molecular water (at 1 ML coverage) or a cluster of just a few water molecules (at submonolayer coverage) around the OH^--H^+ pair could be sufficient to stabilize this configuration. In this case, features from both adsorbed O_WH^- and molecular H_2O would be visible in MIES spectra. However, we would expect the H_2O induced peaks to be weaker than O_WH^- peaks because O_WH^- species are protruding above the layer of adsorbed water molecules laying essentially flat on the surface [8, 24, 55].

The result that hydroxylation of the MgO(100)-water interface lowers its energy offers a natural explanation for periclase transformation to brucite and dissolution of MgO in water. Low-coordinated surface sites and high-index surface planes need not be involved. Water dissociates not because the aqueous solvent increases the reactivity of the MgO surface. Based on known correlations between reactivity and ionicity of the MgO surface [30], we would expect a hydrated surface to be less reactive than the clean surface due to the slightly increased ionicity of the former. In reality, polarization of the surrounding solvent is responsible for stabilization of charged dissociation products. This result is similar to that found in the case of the NaCl-water interface [22], where solvent effects were found to be more significant in regard to interfacial reactivity than interaction with the ionic crystal surface. It is difficult to judge from our cluster calculations whether a fully hydroxylated MgO(100)-water interface is energetically stable. Answering such a question requires studies on coverage dependence of the interactions in the hydroxyl layer. This is beyond the capabilities of the CECILIA model. Another interesting question is how surface relaxation would affect the hydroxylation process. It is possible to model surface relaxation consistently within the embedded cluster model particularly for ionic crystals. We will discuss this aspect in a future study.

Conclusion

The complexity of chemical processes occurring at the solid-liquid interfaces provides a great challenge to theoretical and computational chemistry. By combining the advantages of the embedded cluster method in studying reactive processes at solid-

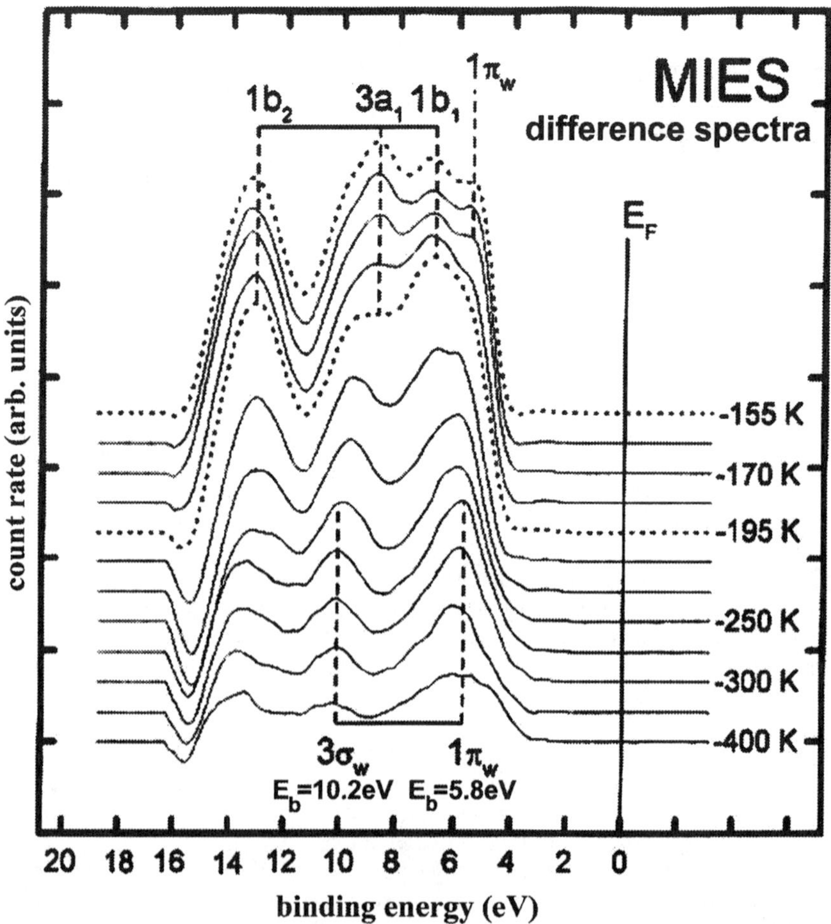

Figure 7: Differential MIES spectra (subtracting the background MgO(100) signal) taken from the MgO(100) surface covered by approximately 3 ML water, as a function of the anneal temperature. (Adapted with permission from ref. [51])

gas interfaces with the dielectric continuum solvation model for modeling reactions in solution, we were able to assemble a unified model called CECILIA. The key advantage of this model is in its ability to model chemical reactions occurring at the solid-liquid interfaces as well as the electronic structure of the interface. We have illustrated it by studying both molecular adsorption and dissociative chemisorption of water at the $MgO(100)$-water interfaces in comparison with recent MD and MC simulations as well as MIES experiments. We have predicted that water is dissociated at the $MgO(100)$-water interface in agreement with the MIES observations. More work is certainly needed in further improving the accuracy of the model, however, progress so far on the CECILIA model is very encouraging. It should be noted that the use of a dielectric continuum solvation model does not provide the liquid structure above the crystal. Recent progress in our lab of using a 3D Reference Interacting Site Model (RISM) showed great promise in providing liquid structure information at a much less computation cost comparing to MD or MC simulations [52].

Acknowledgment

This work is supported by the National Science Foundation.

References

[1] H.-P. Boehm and H. Knozinger, in *Catalysis Science and Technology*, J. R. Anderson and M. Boudart, Editors (Springer-Verlag,1983) p. 39.

[2] D. Ferry, A. Glebov, V. Senz, J. Suzanne, J. P. Toennies and H. Weiss, J. Chem. Phys., 105, (1996), 1697.

[3] M. J. Stirniman, C. Huang, R. S. Smith, S. A. Joyce and B. D. Kay, J. Chem. Phys., 105, (1996), 1295.

[4] C. Xu and D. W. Goodman, Chem. Phys. Lett., 265, (1997), 341.

[5] J. Heidberg, B. Redlich and D. Wetter, Ber. Bunsen-Gessel. Physik., 99, (1995), 1333.

[6] C. A. Scamehorn, A. C. Hess and M. I. McCarthy, J. Chem. Phys., 99, (1993), 2786.

[7] M. R. Chacon-Taylor and M. I. McCarthy, J. Phys. Chem., 100, (1996), 7610.

[8] M. I. McCarthy, G. K. Schenter, C. A. Scamehorn and J. B. Nicholas, J. Phys. Chem., 100, (1996), 16989.

[9] N. H. de Leeuw, G. W. Watson and S. C. Parker, J. Phys. Chem., 99, (1995), 17219.

[10] V. A. Tikhomirov, G. Geudtner and K. Jug, J. Phys. Chem. B, 101, (1997), 10398.

[11] Y. Kuroda, E. Yasugi, H. Aoi, K. Miura and T. Morimoto, J. Chem. Soc., Faraday Trans. 1, 84, (1988), 2421.

[12] S. Coluccia, L. Marchese, S. Lavagnino and M. Anpo, Spectrochim. Acta A, 43, (1987), 1573.

[13] H. Onishi, C. Egawa, T. Aruga and Y. Iwasawa, Surf. Sci., 191, (1987), 479.

[14] A. L. Almeida, J. B. L. Martins, C. A. Taft, E. Longo and W. A. Lester, Jr., J. Chem. Phys., 109, (1998), 3671.

[15] J. Goniakowski and C. Noguera, Surf. Sci., 330, (1995), 337.

[16] C. A. Scamehorn, N. M. Harrison and M. I. McCarthy, J. Chem. Phys., 101, (1994), 1547.

[17] K. Refson, R. A. Wogelius, D. G. Fraser, M. C. Payne, M. H. Lee and V. Milman, Phys. Rev. B, 52, (1995), 10823.

[18] R. L. Segall, R. S. C. Smart and P. S. Turner, in *Surface and Near-Surface Chemistry of Oxide Materials*, J. Nowotny and L.-C. Dufour, Editors (Elsevier,Amsterdam, 1988) .

[19] M. Komiyama and M. Gu, Appl. Surf. Sci., 120, (1997), 125.

[20] M. A. Blesa, P. J. Morando and A. E. Regazzoni, Chemical Dissolution of Metal Oxides. (CRC Press,Boca Raton, 1994).

[21] X.-L. Zhou and J. P. Cowin, J. Phys. Chem., 100, (1996), 1055.

[22] E. V. Stefanovich and T. N. Truong, J. Chem. Phys., 106, (1997), 7700.

[23] E. V. Stefanovich and T. N. Truong, in *Combined quantum mechanical and molecular mechanical methods*, J. Gao and M. A. Thompson, Editors (ACS Books,Washington, DC, 1998) p. 92.

[24] M. A. Johnson, E. V. Stefanovich and T. N. Truong, J. Phys. Chem. B, 102, (1998), 6391.

[25] M. A. Johnson, E. V. Stefanovich and T. N. Truong, J. Phys. Chem. B, 103, (1999), 3391.

[26] T. A. Kaplan and S. D. Mahanti, eds. *Electronic properties of solids using cluster methods*. Fundamental materials research, ed. M. F. Thorpe (Plenum, New York, 1995).

[27] G. Pacchioni, P. S. Bagus and F. Parmigiani, eds. *Cluster Models for Surface and Bulk Phenomena*. (Plenum, New York, 1992).

[28] R. W. Grimes, C. R. A. Catlow and A. L. Shluger, eds. *Quantum mechanical cluster calculations in solid state studies*. (World Scientific, Singapore, 1992).

[29] M. A. Johnson, E. V. Stefanovich and T. N. Truong, J. Phys. Chem. B, 101, (1997), 3196.

[30] E. V. Stefanovich and T. N. Truong, J. Chem. Phys., 102, (1995), 5071.

[31] E. V. Stefanovich and T. N. Truong, J. Phys. Chem.B, 102, (1998), 3018.

[32] J. Tomasi and M. Persico, Chem. Rev., 94, (1994), 2027.

[33] C. J. Cramer and D. G. Truhlar, in *Reviews in Computational Chemistry*, K. B. Lipkowitz and D. B. Boyd, Editors (VCH Publishers,New York, 1994) p.1.

[34] T. N. Truong and E. V. Stefanovich, Chem. Phys. Lett., 240, (1995), 253.

[35] T. N. Truong and E. V. Stefanovich, J. Chem. Phys., 109, (1995), 3709.

[36] T. N. Truong, U. N. Nguyen and E. V. Stefanovich, Int. J. Quant. Chem.: Quant. Chem. Symp., 30, (1996), 403.

[37] T. N. Truong, Int. Rev. Phys. Chem., 17, (1998), 525.

[38] A. Klamt and G. Schüürmann, J. Chem. Soc., Perkin Trans. II, (1993), 799.

[39] F. M. Floris, J. Tomasi and J. L. Pascual-Ahuir, J. Comp. Chem., 12, (1991), 784.

[40] W. L. Jorgensen and J. Tirado-Rives, J. Am. Chem. Soc., 110, (1988), 1657.

[41] R. A. Pierotti, Chem. Rev., 76, (1976), 717.

[42] M. J. Huron and P. Claverie, J. Phys. Chem., 76, (1972), 2123.

[43] W. Stevens, H. Basch and J. Krauss, J. Chem. Phys., 81, (1984), 6026.

[44] J. L. Pascual-Ahuir, E. Silla and I. Tuñon, J. Comp. Chem., 15, (1994), 1127.

[45] E. V. Stefanovich and T. N. Truong, Chem. Phys. Lett., 244, (1995), 65.

[46] J. Günster, G. Liu, V. Kempter and D. W. Goodman, J. Vac. Sci. Technol. A, 16, (1998), 996.

[47] D. Ochs, W. Maus-Friedrichs, M. Brause, J. Günster, V. Kempter, V. Puchin, A. Shluger and L. Kantorovich, Surf. Sci., 365, (1996), 557.

[48] X. D. Peng and M. A. Barteau, Surf. Sci., 233, (1990), 283.

[49] X. D. Peng and M. A. Barteau, Langmuir, 7, (1991), 1426.

[50] M. J. Frisch, G. W. Trucks, H. B. Schlegel, P. M. W. Gill, B. G. Johnson, M. W. Wong, J. B. Foresman, M. A. Robb, M. Head-Gordon, E. S. Replogle, R. Gomperts, J. L. Andres, K. Raghavachari, J. S. Binkley, C. Gonzalez, R. L. Martin, D. J. Fox, D. J. Defrees, J. Baker, J. J. P. Stewart and J. A. Pople, (Gaussian, Inc., Pittsburgh, PA, 1993.

[51] M. A. Johnson, E. V. Stefanovich, T. N. Truong, J. Günster and D. W. Goodman, J. Phys. Chem. B, 103, (1999), 3391.

[52] V. Shapovalov, T. N. Truong, A. Kovalenko and F. Hirata, Chem. Phys. Lett., 320 (2000), 186.

[53] S. Russo and C. Noguera, Surf. Sci., 262, (1992), 245.

[54] J. Goniakowski, S. Bouette-Russo and C. Noguera, Surf. Sci., 284, (1993), 315.

[55] A. Marmier, P. N. M. Hoang, S. Picaud, C. Girardet and R. M. Lynden-Bell, J. Chem. Phys., 109, (1998), 3245.

Chapter 11

Modeling of Semiconductor–Electrolyte Interfaces with Tight-Binding Molecular Dynamics

P. K. Schelling[1] and J. W. Halley[2]

[1]Argonne National Laboratory, 9700 South Cass Avenue,
Argonne, IL 60439
[2]School of Physics and Astronomy, University of Minnesota,
Minneapolis, MN 55455

To understand such phenomena as ion and electron transfer at the semiconductor-electrolyte interface, it is important to model the atomic and electronic dynamics of the electrode. For many such problems, first-principles techniques are too computationally costly. We have developed a self-consistent tight-binding model to address these questions related to the electrode-electrolyte interface. The model has been parameterized by a fit to first-principles results for a bulk oxide. The resulting model has been shown to be useful for describing surface and defect structures. Since our model has been developed to have a computational cost which scales linearly with the number of atoms, simulation of systems with 1000 ore more ions is now within reach. We will discuss application to metal-oxide interfaces.

Introduction

Reactions at the electrode-electrolyte interface are key to many technological subjects such as batteries, fuel cells, and chemical etching. While first-principles methods are able to give reasonable results for many ground state properties of oxides, they remain limited to rather small supercells due to computational constraints. For many problems which require large scale molecular dynamics simulation, it seems likely that first-principles techniques will not be practical in the near future. However, it is possible to use first-principles calculations to parameterize simpler models which still capture the basics physics. This chapter describes work done on innovative tight-binding models which have been extended beyond most similar techniques by including electron-electron interactions. Our methods represent a way to extend first-principles techniques to model phenomena on larger length and time scales.

Tight Binding Model For Rutile

The formal approach used for the Coulomb self-consistent tight binding model is described elsewhere (1). We start with the formal expression which contains terms to describe electron kinetic energy as well as electron-electron, electron-ion, and ion-ion interactions. Coulomb interactions are treated by describing the ions and the electron density as point charges with multipole moments. The single particle wave functions $\psi_\lambda(\vec{r})$ are taken to be a linear combination of nonorthogonal basis function as

$$\psi_\lambda(\vec{r}) = \sum_{iv} c_{iv,\lambda} \phi_{iv}(\vec{r})$$

(1)

The $\phi_{iv}(\vec{r})$ is a nonorthogonal orbital v located at site i. The $c_{iv,\lambda}$ represent the expansion coefficients in this basis for the eigenfunction λ. The wave function can also be expressed in terms of an orthogonal basis

$$\psi_\lambda(\vec{r}) = \sum_{iv} \tilde{c}_{iv,\lambda} \tilde{\phi}_{iv}(\vec{r})$$

(2)

Here the $\tilde{c}_{k\mu,\lambda}$ are the expansion coefficients for the orthogonal basis states. The basis states themselves are related by

$$\tilde{\phi}_{iv}(\vec{r}) = \sum_{k\mu} \phi_{k\mu}(\vec{r}) S_{k\mu,iv}^{-1/2} \; .$$

(3)

where the S matrix is the overlap matrix. The total energy expression use is, in terms of the orthogonal basis

$$E_{tot} = \sum_i E_i(\{Q_{i,s}\},\{R\}) + \sum_{ijvv'} Q_{iv,jv'} \left(\delta_{i,j} \tilde{v}_{iv,jv}^{(1)} + (1 - \delta_{i,j}) \tilde{t}_{iv,jv} \right) +$$

$$(1/2) \sum_{i \neq j} \frac{e^2 (Z_i - Q_i)(Z_j - Q_j)}{R_{ij}} - \sum_{\lambda iv} \varepsilon_\lambda n_\lambda (\tilde{c}_{iv,\lambda}^* \tilde{c}_{iv,\lambda} - 1) \; .$$

(4)

The first term describes onsite interactions, the second term describes covalent bonding, and the third term describes long range Coulomb interactions. The last term is added to assure orthonormality of the single particle wave functions, with ε_λ being a Lagrange multiplier. We assume that the intrasite terms contribute an energy of the form $Q_{i,s} = \sum_{v \in s} Q_{iv,iv}$ where s is a 'shell' of one electron states which are degenerate in the atom or ion, and R describes the coordinates of the ions in the system. In other words, we assume that the energy of a site depends only on the number of electrons on the site and also on the locations of neighboring ions. The term describing covalent bonding and multipole interactions involves the Q matrix defined in terms of the expansion coefficients as

$$Q_{iv,jv'} = \sum_{\lambda} n_{\lambda} \tilde{c}^{*}_{iv,\lambda} \tilde{c}_{jv',\lambda}$$

(5)

where n_{λ} is the number of occupied eigenstates with index λ. The diagonal elements of this matrix describe the local charge Q_j on an atomic site. The long- range Coulomb interactions are treated within the Hartree approximation, with the local atomic charges described by Z_j-Q_j, where Z_j is the core charge.

By varying the coefficients $c_{iv,\lambda}$ we obtain the single particle equation which we actually solve. The single particle equation is solved to obtain a new set of coefficients $c_{iv,\lambda}$, from which the site charges Q_j may be obtained. Given the new set of charges, a new single particle equation is found. We again solve the single particle equation, and repeat the entire procedure until self consistency is obtained.

The model described above needs a functional form for the hopping integrals $t_{i\mu,jv}$ to describe how these parameters depend on ion separation. We assume that the dependence on separation between ions is $t_{i\mu,jv}=t_0(R_0/R_{ij})^n$ where R_{ij} is the separation between the ions. The parameters t_0 and n were found by fitting the model to first-principles cohesive energies and band structure for a set of lattice parameters[2]. Since the model can describe the energetics as a function of ion separation, we are able to use the model in relaxational as well as molecular dynamics studies.

Atomic and Electronic Structure of Rutile Surfaces and Grain Boundaries

As a first test of the tight-binding model, we studied the structure of rutile surfaces. This work is described in reference 2. Although the model as described above does well for bulk properties of rutile, there is no guarantee that it will work well in an environment very different from where it was parameterized. Thus, the ability to predict properties of rutile surface represents an important test of the model. Furthermore, a good description of rutile surfaces is essential before addressing problems related to the solid/liquid interface. We found very good agreement with surface structure and energetics, both with first-principles and experimental results. Significant charge transfer was found on the surface, indicating the importance of including effects of electron-electron interactions.

In addition to the study of surfaces, we also simulated a $\Sigma 5$ 001 twist grain boundary. We studied systems containing 300 and 540 ions. The relaxed structure for the 540 ion grain boundary is shown in Figure 1. Periodic boundary conditions were applied in all three dimensions, and consequently two identical twist grain boundaries are present in the simulation. Grain boundaries often dominate material properties in technological applications. For example, grain boundaries interfaces may be associated with localized electronic states which will affect transport properties by trapping carriers. In addition, it is well known that grain boundaries can act as a source or sink for point defects. Both of these effects may have important

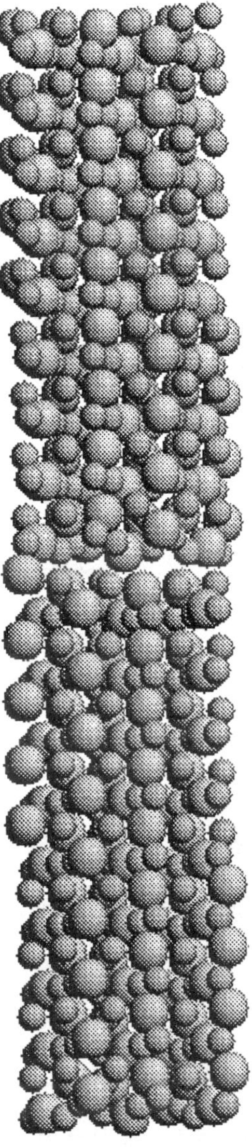

1. 540 ion Σ5 twist grain boundary after relaxation using tight binding

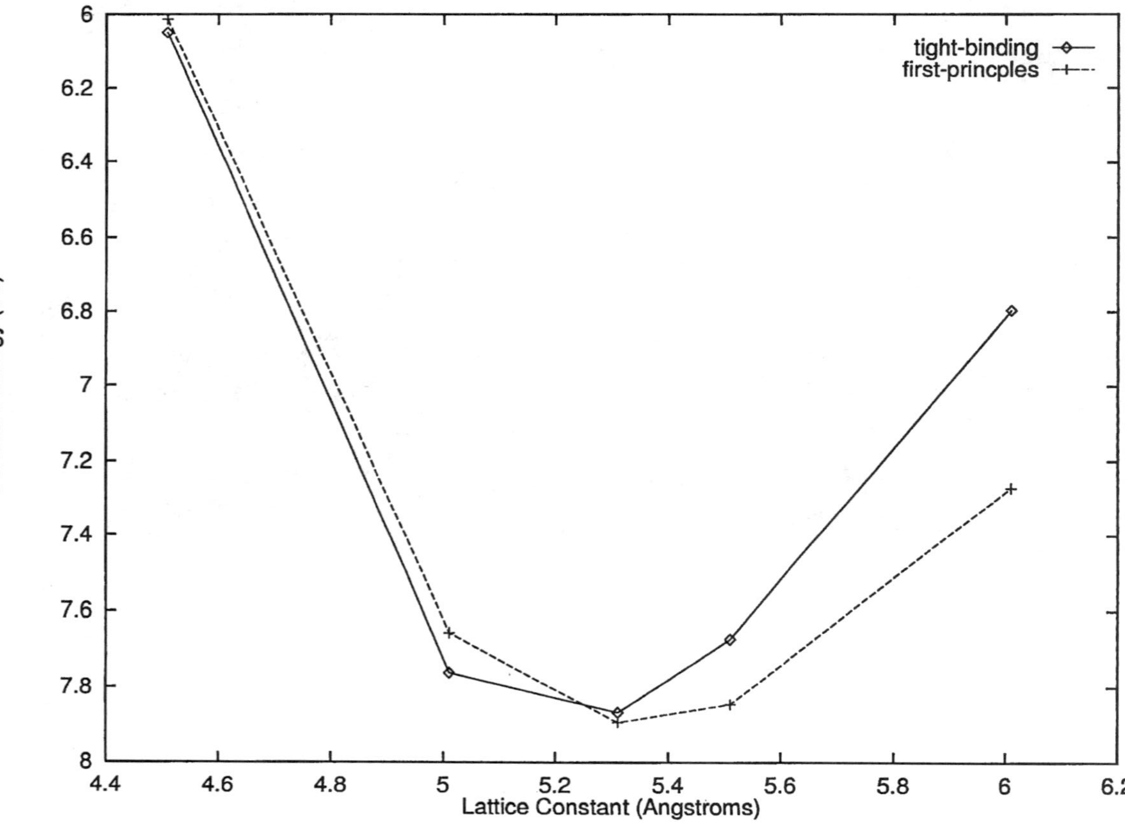

2. Ti cohesive energy per ion for the first-principles and tight binding model.

implications in electrochemical problems such as electron transfer reactions or passivation.

As a first step, a classical model, using only pair potentials and point charges, was used to relax the grain boundary. The starting system was taken to be two perfect 001 slabs which were rotated to obtain the desired grain boundary structure. In general, the resulting structure will depend on the starting system used, and it may be important to perform finite temperature annealing in order to obtain a ground state relaxed structure. For the relaxation performed here, this was not done, so it is entirely possible that the relaxed structures obtained do not represent a ground state relaxed structure. The final relaxation was done using the tight binding model. We found significant expansion along the 001 axis, and a rather large grain boundary energy. The grain boundary energy is defined as

$$\sigma_{gb} = \frac{1}{2A}\left[E_{gb} - nE_{bulk}\right] \tag{6}$$

Here, A is the sample area in the grain boundary plane, E_{gb} is the total energy of the grain boundary, n is the number of unit cells present in a perfect bulk crystal with the same number of ions as the grain boundary sample, and E_{bulk} is the total energy per unit cell for the perfect bulk crystal. The factor of 1/2 accounts for the presence of two identical grain boundaries present in the simulation. We find for the 300 ion system   and for the 540 ion system. By comparing these results with the energy per area of a 001 surface we may obtain the amount of energy per area needed to break the grain boundary into two slabs with two 001 surfaces each. We can use

$$\gamma = 2\sigma_{001} - \sigma_{gb} \tag{7}$$

to obtain the energy per area to cleave one of the grain boundaries. The grain boundary volume expansion was found to be quite significant. For the 300 ion system, the c axis expansion was 3.2 A, and for the 540 ion system is was 1.6 A. This large difference between the 300 and 540 ion systmes we believe is due to significant interaction between the two grain boundaries in the simulation for the 300 ion system. For the 300 ion case, the entire system is about 33 A in length, making each slab about 16.5 A long, which is 5 unit cells in thickness. However, since two unit cells are directly in the grain boundary region, there are only three unit cells in between. To fully resolve this issue, more simulations need to be performed to establish at what point the grain boundary energy and volume expansion are well converged. In both systems, the grain boundary expansion is quite significant. It is possible that a real twist grain boundary in this system may also include a certain number of intrinsic point defects, which will lower the energy and also decrease the grain boundary expansion.

We found, as with our previous calculations for surfaces, that the electronic density of states was very similar to a bulk sample. This indicates that these grain boundaries may not be very important for some electronic properties, such as electron-hole generation or recombination. We did, however, find localized electronic states associated with the grain boundary region which are expected to act as scattering centers for conduction electrons.

Metal-Oxide Interfaces

The layer of oxide usually present on metals is very important in the context of electrochemical problems. To begin to address problems related to passivation, a description of the metal-oxide interface is required. As a first step, we have begun to develop self-consistent tight-binding models which can describe transition metals. This requires use of a larger basis set to describe the delocalized metallic wave functions. We use a basis comprised of 3d, 4s, and 4p titanium orbitals. Applying the same techniques of fitting to first-principles calculations of the total energy and band structure, we are able to obtain a model which reproduce the first-principles cohesive energy. We found that the titanium metal was stable with a lattice constant of 5.3 A. Site energies in this model were taken to be a polynomial function of the local charge Q as $E(Q) = E_0 + E_1(\{R\})Q + (1/2)E_2Q^2$. We treated the term linear in Q as a function of the separation between titanium ions R as

$$E_1(R) = A_s\left[1 + B_s\left(\frac{1}{R}\right)^n\right] \tag{8}$$

where R is the ion separation and n and B_s are fitting parameters. The A_s was chosen so that the orbital energies reproduce the orbital energies for the atomic shell s for the isolated ion. Hopping integrals were taken to scale with distance as $(1/R)^n$, and fermi function cutoffs were used to smoothly cut off the interactions at 5.55 A.

With a model for titanium in hand, it must be incorporated into the existing model of TiO_2, which only included 3d states in the parameterization. We found that the model worked reasonably well if we included a term to describe the fact that the 4s and 4p orbitals lie above the 3d orbitals in TiO_2. Therefore, the site energies for the 4s and 4p orbitals were taken to be functions of the local environment of the oxygen. The 3d states did not require any such term.

As a first test of this model, we attempted to relax an interface of pure titanium with a thin layer of oxide. In Figure 3a and 3b we show results of a first attempt at such a calculation. In this calculation we have place 7 atomic layers of fcc titanium between two sheets of a total of 9 layers of rutile TiO_2 and allowed the atomic positions to relax in response to forces computed self consistently at each relaxation step. Periodic boundary conditions were applied in all three dimensions. Approximately 500 steps were needed to sufficiently relax the structure. There are 36 units of TiO_2 and then 56 titanium ions in the fcc metal, for a total of 164 ions. The cell is fixed to be 9.2 A in the 100 and 010 plane, and 36 A in the 001 direction. The starting structure is shown if Figure 3a viewed along the 110 direction. In the final relaxed structure shown in Figure 3b, the oxygen ions on the top and bottom of the slab are 24.26 A apart, resulting in a vacuum space of 11.74 A.

The interesting result of Figure 3b is that the titanium ions which were originally in the 001 face of the TiO_2 which is in contact with the metal layer have relaxed significantly into the metal to form strong bonds with the titanium ions in the metal. These ions still form strong bonds with the oxygen ions a well, and have a large positive charge of 2.09 electron charges. The charges on the metal ions are significant, ranging from between 2.09 and −1.26 electrons. It is revealing to look at Figure 4 which shows the sum of charges in a plane as a function of distance across the structure. There is a net negative charge in the metallic region of about −3.3 electrons if one does not count the titanium ions which relaxed in from the oxide layer. It is clear that in order to fully describe the interface, a larger metal region is needed as one expects the oscillating charges to eventually approach zero in the bulk. Charge oscillations in the metal region were also reported by R. Benedek for simulations of an Mgo/Cu interface (3). It is also interesting to note that he found lower adhesive energies for nonpolar oxide interfaces with metal. The 001 surface of TiO_2 is comprised of neutral planes making it a nonpolar interface. However, after relaxation as described above, the interface becomes very polar.

O(N) Method

All of the simulations described above were performed by exact diagonalization to obtain the single particle wave functions. The problem of diagonalizing a matrix to obtain the eigenvectors has a computational cost which in principle scales as N^3 where N is the number of basis functions. In practice we find that because the tight binding matrix is sparse the computational cost scales more favorably, approximately as $N^{2.5}$. This undesirable scaling makes it impossible to extend tight-binding models to systems of even 1000 ions when direct diagonalization is used.

However, recently many researchers have developed techniques which are capable of reducing the problem of computing the electronic structure to a calculation of the density matrix, whose elements $Q_{i\mu,j\nu}$ were defined in equation(5). As one can see from the total energy expression defined by equation (4), the density matrix alone is adequate to obtain energy and hence forces. Because of the limited range associated with the hopping integrals, we only need the Q matrix within a limited range.

We start by defining an energy E which is equivalent to the sum of eigenvalues for the occupied states

$$E = tr[QH] = \sum_{i\mu,j\nu} Q_{i\mu,j\nu} H_{j\nu,i\mu} \quad . \tag{9}$$

In this equation, H is the single particle Hamiltonian. The total number of electrons is just

$$N = tr[Q] = \sum_{i\mu} Q_{i\mu,i\mu} \quad . \tag{10}$$

At first it might seem that it would be possible to directly minimize E defined by equation (9) with respect to the density matrix. This does not work, however, because

3a. Atomic positions of metal-oxide interface before relaxation.

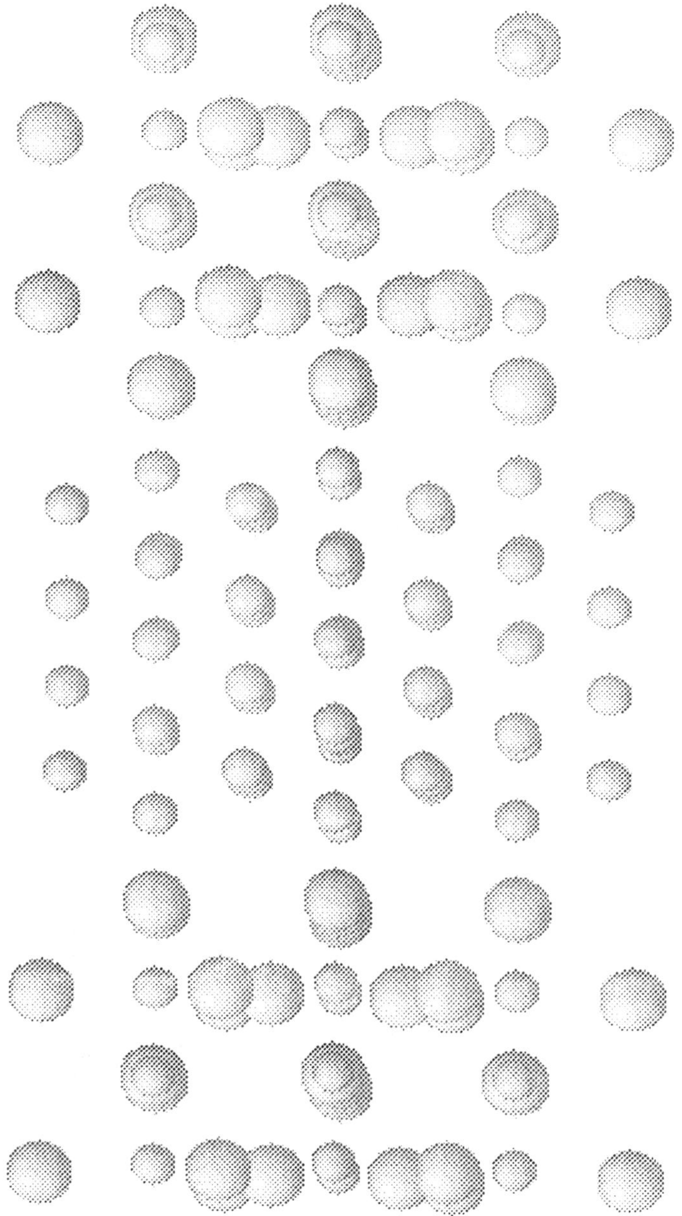

3b. Relaxed metal-oxide interface.

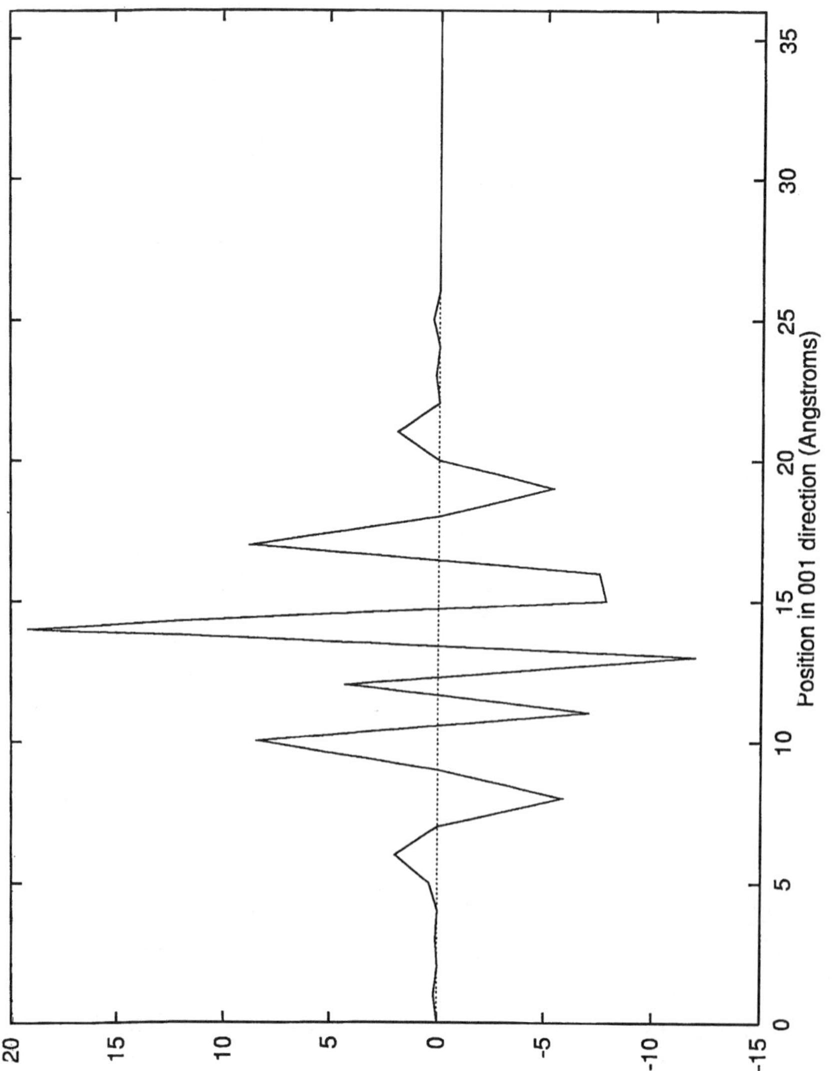

4. Charges summed over layers as a function of z (along 001 directions)

of the requirement that the density matrix is idempotent, which means that $Q^2=Q$. This requirement is equivalent to the requirement that the underlying eigenstates be orthogonal. The technique we applied to our tight binding model was first developed by Li, Nunes, and Vanderbilt (4). This technique was first inspired by the realization that if a matrix Q is nearly idempotent, the matrix $\tilde{Q} = 3Q^2 - 2Q^3$ will be more nearly idempotent (5). This is commonly referred to as a purity transformation. Li *et al* make use of trick and minimize

$$\Omega = tr[(3Q^2 - 2Q^3)(H - \mu)] \tag{11}$$

where μ is a chemical potential added to conserve the number of electrons N. It was also shown by Nunes and Vanderbilt (6) that for a nonorthogonal basis, the above expression is replaced by,

$$\Omega = tr[(3QSQ - 2QSQSQ)(H - \mu S)] \tag{12}$$

where S is the overlap matrix and the Q matrix is defined in the same was as it was before.

The O(N) trick enters because the Q matrix is predicted to be a banded matrix. In the case of an insulator, the $Q_{i\mu,jv}$ falls off exponentially with R_{ij}, and for metals, it is expected to fall off as R^{-d}, where d is the dimensionality of the system (7). Therefore, one can neglect elements of the Q matrix outside of some radius R_c.

For the model described previously, the charges which are treated self-consistently are defined within an orthogonal basis. For example, the charge on a site is defined as

$$Q_i = \sum_{\lambda\mu} n_\lambda \tilde{c}_{i\mu,\lambda}^* \tilde{c}_{i\mu,\lambda} \tag{13}$$

But the coefficients thus defined in the orthogonal basis must be found from the coefficients in the nonorthogonal basis as

$$\tilde{c}_{i\mu,\lambda} = \sum_{jv} S_{i\mu,jv}^{+1/2} c_{jv,\lambda} \ . \tag{14}$$

This means that even with the O(N) method as described above, the S matrix would still need to be diagonalized to obtain $S^{1/2}$, which in principle entail $O(N^3)$ operations. We therefore adopt a different definition of the local charge on a site as

$$Q_i = \left(\tfrac{1}{2}\right)\sum_\lambda n_\lambda \sum_{\mu,jv} [c_{i\mu,\lambda}^* S_{i\mu,jv} c_{jv,\lambda} + c_{jv,\lambda}^* S_{jv,i\mu} c_{i\mu,\lambda}] \tag{15}$$

so that the charges can be defined by using the S matrix without having to diagonalize it. Using this new definition of the local charges, we obtain a different expression for the single particle Hamiltonian. In principle, since this tight binding formulation is different from the one described in the previous section, it needs to be reparameterized. However, our results indicate that the charges and energies calculated have not changed very significantly from the previous model.

To use this O(N) method to minimize the total energy we must first start with a guess for the Q matrix, which then provides the point charges. Then the single particle Hamiltonian is computed. We then minimize Ω defined by equation (12) with respect to the Q matrix by computing gradients which are followed to the minimum by a steepest descents algorithm. Given the new Q matrix, we can compute a new set of charges, and therefore a new Hamiltonian, and the whole process is iterated until self consistency is achieved.

For TiO_2 we studied the errors in the band energy E defined as the sum of eigenvalues, charges, and the total electron number as a function of cutoff radius R_c. Our results showed that 6 A to 7 A is a reasonable cutoff for the system. We have begun to study strained and defective rutile surfaces with this new technique. We should eventually be able to extend these simulations to 1000 ions or more, and since the matrix multiplication is very local, it should work very well on parallel computers.

Conclusions

We have demonstrated the usefulness of tight binding models in bridging between first-principles simulation and larger scale molecular dynamics studies. As the tight binding method seems to be useful for describing metals as well as oxides, we have begun to describe the interface of a metal with an oxide, with a preliminary calculation giving interesting results suggesting that the oxide prefers to form a polar interface. Further work will be directed towards refinement of the model, and the extension to more realistic interfaces. We have begun to further address the computational cost of these methods by using O(N) methods to calculate the electronic structure part of the problem. This, along with the work already done, should allow us to address problems such as passivation. However, it is likely that tight-binding models describing the liquid side of the interface need to be constructed.

Acknowledgements

This work was supported by US Department of Energy, Division of Material Sciences, Office of Basic Energy Sciences under grant DE-FG02-91-ER45455 and by the Minnesota Supercomputing Institute.

References

1. N. Yu and J. W. Halley, Phys. Rev B **51**, 4768 (1995)
2. P.K. Schelling and J.W. Halley, Phys. Rev B **58**, 1279 (1998)

3. R. Benedek, Phys. Rev. B **54**, 7696 (1999)
4. X.-P. Li, R.W. Nunes and D.Vanderbilt, Phys Rev. B **47**, 10891 (1993)
5. R. Mcweeney, Rev. Mod. Phys. **32**,335 (1960)
6. R.W. Nunes and D. Vanderbilt, Phys. Rev. B **50**, 17611 (1994)
7. W. Kohn, Phys. Rev. **115**, 809 (1959)

Organic Liquid–Solid Interfaces

Structural and Dynamic Properties of Hexadecane Lubricants under Shear Flow in a Confined Geometry

Yanhua Zhou[1], Tahir Cagin[1], Elaine S. Yamaguchi[2], Andrew Ho[2], Rawls Frazier[2], Yongchun Tang[2], and William A. Goddard III[1,*]

[1]Materials and Process Simulation Center, Beckman Institute (139–74), Division of Chemistry and Chemical Engineering, California Institute of Technology, Pasadena, CA 91125
[2]Chevron Chemical Company, Oronite Global Technology, 100 Chevron Way, Richmond, CA 94802

Using shear dynamics simulations we investigated the structure and dynamics of hexadecane (n-$C_{16}H_{34}$) lubricant films of a nanoscale thickness, confined between two solid surfaces (Fe_2O_3) covered with a self-assembled monolayer of wear inhibitors [i.e., dithiophosphate molecules $DTP = S_2P(OR)_2$ with $R = iPr$, iBu, and Ph]. We found significant density oscillations in the lubricant films, especially near the top and bottom boundaries. From the density oscillations we can define 9-10 layers for a film of 44 Å thickness, and 5 layers for a 20 Å thick film. The motions of individual lubricant molecules in the direction perpendicular to the surfaces are rather restricted, spanning only 1-2 layers during the entire 200 ps. We also observed the stick-slip motion of the lubricant molecules near the bottom and top boundaries in the direction of shear. However, the change from stick to slip state (or *vice versa*) for a lubricant molecule does not correlate with the change in its radius of gyration or end-to-end distance. The characteristics of the stick-slip motion of the lubricant molecules are strongly influenced by the type of organic R-group in the wear inhibitor molecules.

Advances in experimental and computer simulation techniques have led to an increasing understanding of fluids in thin films at the atomic scale *(1,2)*. It has been found that when a liquid is confined by solid surfaces, its physical properties change at first gradually, then dramatically, as the separation becomes progressively narrower *(3,4)*. For ultra thin films of 10 Å thick or less, a number of solidlike behaviors have been found including capability of resisting shear forces *(5,6)*. In this paper we focus on hexadecane films of around 40 Å thick and investigate their behavior and dynamics in response to a shear flow. The primary objective of the study is to understand how lubricants with nanoscale thickness behave when they pass engine surfaces and to assist in designing engine surfaces with reduced engine wear *(7,8,9)*.

The model of the present study builds upon our previous static studies of iron-oxide surfaces covered by a self-assembled monolayer (SAM) of adsorbed wear inhibitors ("brush") *(10)*. A hexadecane lubricant film is sandwiched between two such protected iron-oxide surfaces. A planar flow is generated by moving the top surface and keeping the bottom surface stationary. Our model system consists of realistic lubricants, realistic solid surfaces, and realistic wear inhibitor molecules adsorbed on the surfaces.

The wear inhibitor molecules are dithiophosphates [$DTP=S_2P(OR)_2$, where R represents one of the three types of organic groups: R =isopropyl (iPr), isobutyl (iBu), and phenyl (Ph)].

Simulation Model

We have described the the simulation details performed in this study in a recent paper *(11)*. A typical structure of the system is shown in Figure 1. The fluid consisting of hexadecane molecules was confined between two iron-oxide surfaces, each of which is covered by 8 dithiophosphate (DTP) molecules (acting as wear inhibitors). The Fe_2O_3 surface is oriented in such a way that the (0001) surface is perpendicular to the z-axis. A planar flow is generated in the xz-plane by moving the top surface at a constant velocity along the x-axis while keeping the bottom surface fixed.

The initial velocities were chosen from a Maxwell-Boltzmann distribution at $T = 500$ K. We added a constant velocity, v, to all atoms of the top slab of Fe_2O_3 and to the top SAM layer. The lubricant and bottom layer started with the initial set of velocities from Maxwell-Boltzmann distribution with zero center of mass velocity. In the simulations, the iron-oxide surfaces are sheared uniformly with respect to each other at a constant rate and the DTP plus lubricant are allowed to readjust to the shearing motion according to the forces acting on them. We carried out ~200 ps of shear dynamics simulations. The time step is 1 fs. The trajectories were stored every 0.05 ps. The force field (FF) used was obtained by combining quantum mechanics calculations on Fe-DTP clusters with the Dreiding force field *(12)* to describe the DTP and iron-oxide interactions. The derivation and the parameters for this FF are described elsewhere *(8,9)*.

The results presented below are from four different simulations:

1. $R = $ iPr, $v = 1$ Å/ps, and the lubricant thickness H = 45 Å,

2. $R = $ iBu, $v = 1$ Å/ps, and H = 45 Å,

3. $R = $ Ph, $v = 1$ Å/ps, and H = 43 Å, and

4. $R = $ iPr, $v = 0.5$ Å/ps, and H = 20 Å.

A list of parameters of these simulations are given in Table I. For each system the cell length in the z direction, L, was determined by minimizing the energy of

Table I. Values of the Geometry Parameters Shown in Figure 1 for Various SAMs.

SAM	L^a(Å)	H^b(Å)	D^c(Å)	h^d(Å)	Run Time(ps)
(a) 44 Å Film					
iPr	83.84	44.99	19.42	6.14	201
iBu	84.89	44.57	20.16	6.71	213
Ph	83.60	43.49	20.06	6.66	203
(b) 20 Å Film					
iPr	63.84	20.30	21.77	6.09	360

[a]Total length of periodic cell in z direction.

[b]Thickness of the lubricant region.

[c]Thickness of the "brush" wall consisting of Fe_2O_3 and DTP monolayers.

[d]Thickness of wear inhibitor SAM.

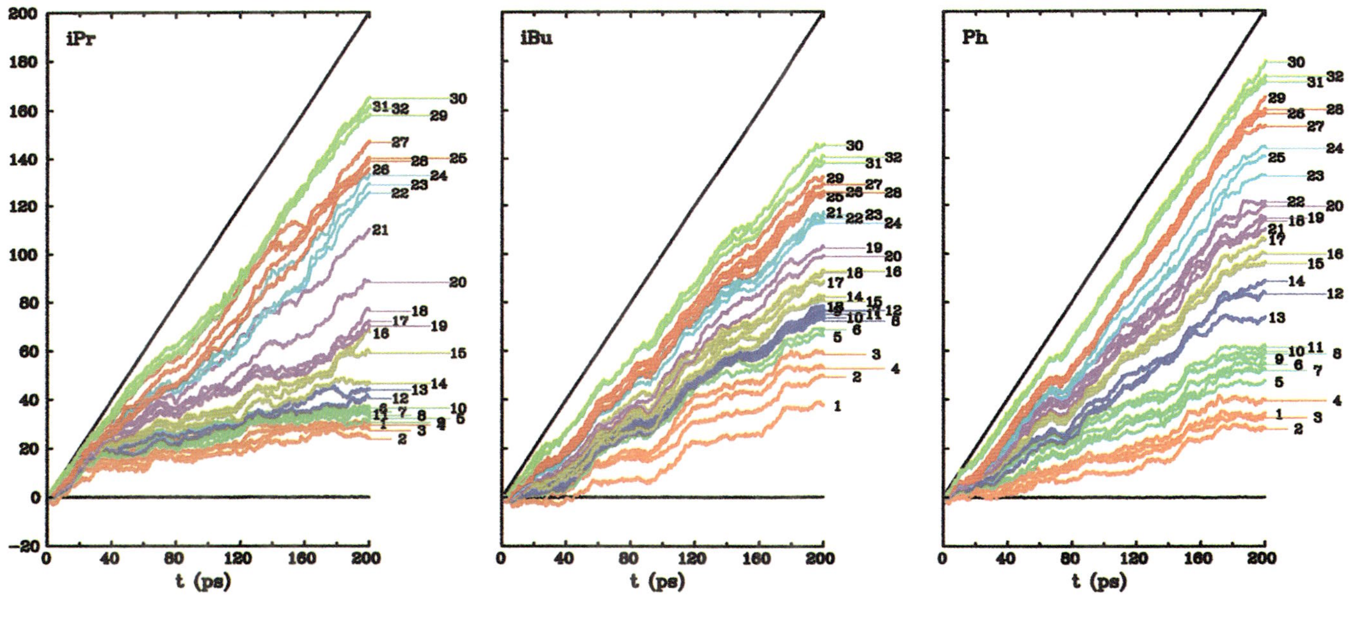

Figure 4. The displacement of each lubricant molecule in the shearing (x) direction as a function of time, for iPr, iBu, and Ph. For easy viewing, the 32 lubricant molecules are partitioned into 8 layers by the final positions in x; each layer is indicated with a distinct color. The label for each molecule is in agreement with number of molecule in Figure 3.

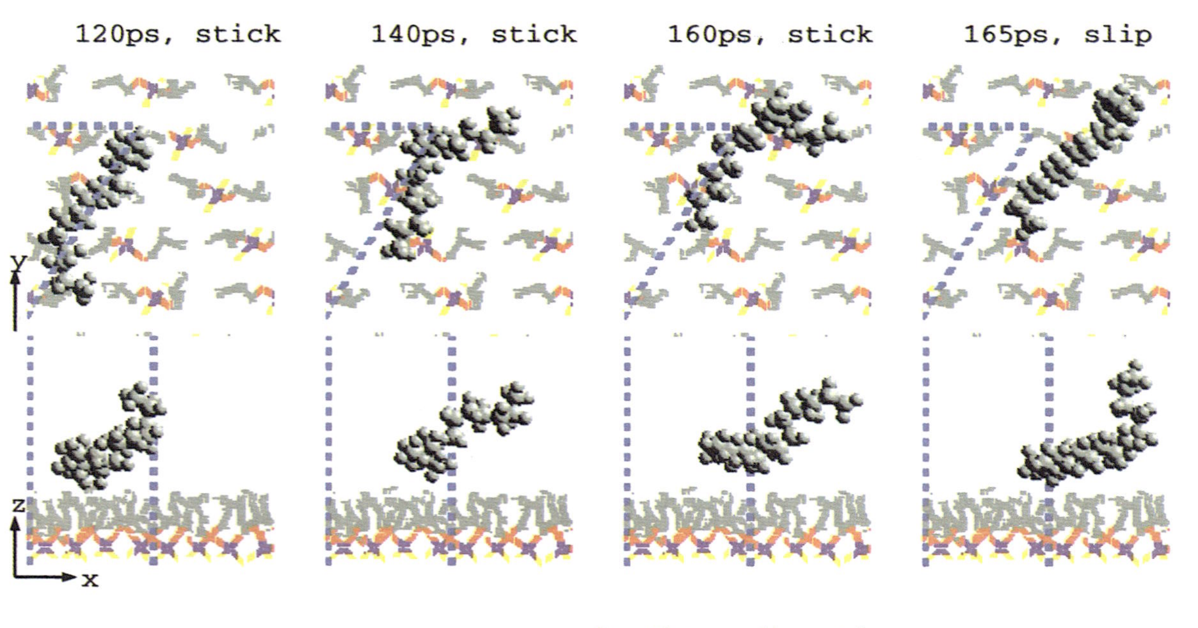
120ps, stick
140ps, stick
160ps, stick
165ps, slip
Y
Z
X
Shear direction

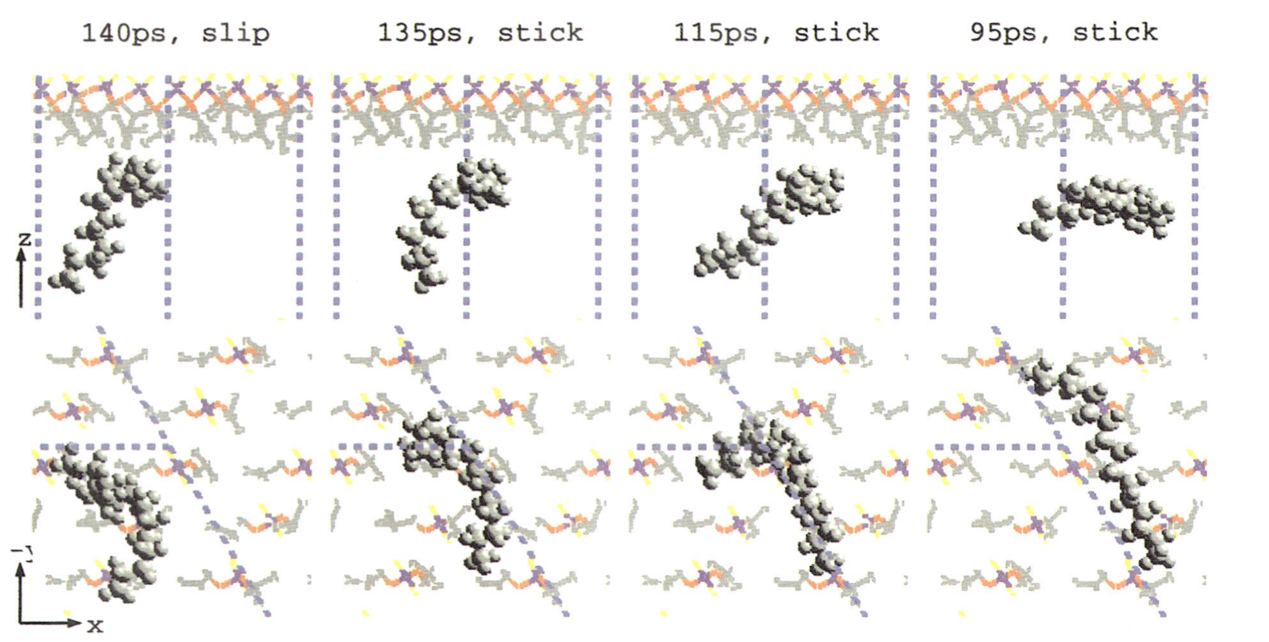

Figure 5. Snapshots of lubricant molecules at various stick-states from the iBu case. (a) Molecule-2 from the bottom: top view – looking down along the z-axis, side view – looking along the y-axis; and (b) molecule-31 from the top: side view – looking along the y-axis, "bottom" view – looking up along the $(-z)$-axis.

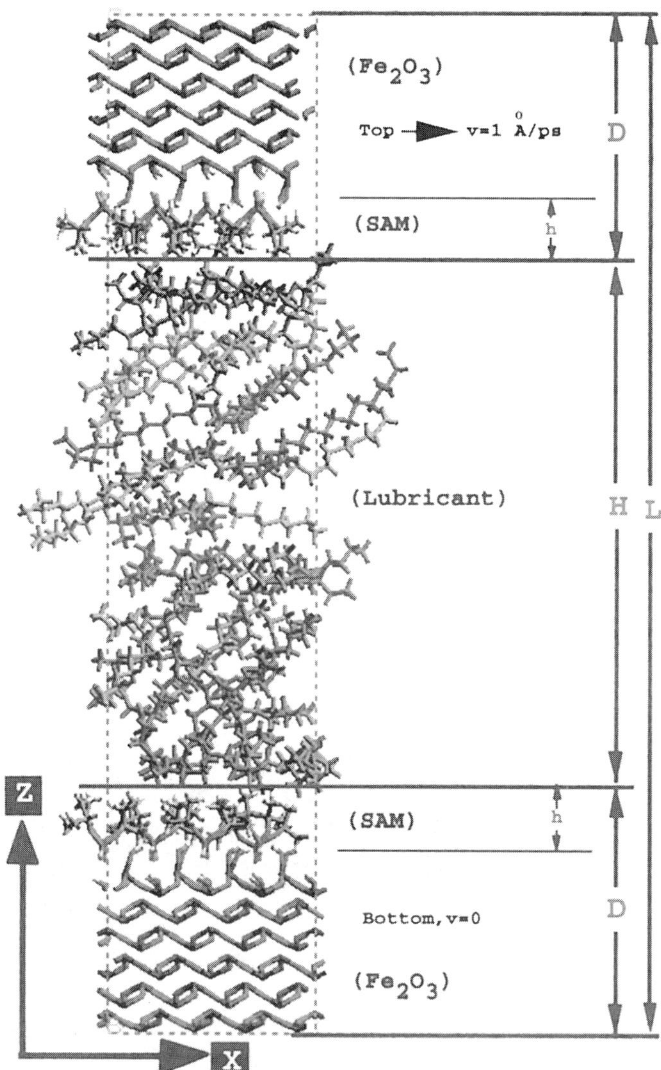

Figure 1. The structure and setup of the system for the shear dynamics simulations. Shown here is the snapshot after 200 ps for the case of iPr. The values for H, L, etc. are in Table 1. For the middle region, the 32 C_{16} molecules are partitioned into 8 different layers according to their z_{cm}; each layer is plotted with a different color.

the entire system (with fixed cell parameters in the other two directions). As a result, the thickness of the 32 $C_{16}H_{34}$ layer is $\sim$44 Å, and that of the 16 $C_{16}H_{34}$ layer is $\sim$20 Å.

Results and Discussions

Density Oscillations

Figure 2 shows the distribution of the density of lubricants along the z-axis from the four simulations. We observed distinct density oscillations in all cases. Very similar results have been found in previous studies *(13,14)*. The density oscillations are significant near the walls (within 15 to 20 Å). In the central region of the fluid, the layering decreases, with the density approaching that of the bulk lubricant. For the separations of $\sim$44 Å we found 9-10 layers in the density profile. For the thinner system with just $\sim$20 Å separation (Figure 2d), we found 5 layers, but here the density oscillations are easily noticeable even in the central region of the film. In an earlier study *(11)* we looked at the distribution of carbon atoms and that of hydrogen atoms separately. There we found that the distribution of hydrogen atoms is quite uniform and that the density oscillations are mostly a result of the oscillation in carbon atoms.

In the present study we further analyzed the contributions from each individual molecule to the oscillatory distribution of total carbon atoms in the lubricant film. The results for iBu are shown in Figure 3 (similar results are obtained for iPr and Ph). Figure 3(a) shows the results obtained by averaging over the second 100 ps of the simulation. In the figure, the carbon atom distribution for each of the 32 molecules (the dark solid curves) is plotted against the distribution of total carbon atoms (the light solid curve). The results clearly indicate that the motions of the molecules in the z-direction are mostly confined within 1-2 layers during the 100 ps of dynamics. Occasionally a molecule can span three layers, such as molecule-10, 18, and 21. Notice that the molecules near the bottom and top interfaces are the most confined, due to stronger interactions with the "brush" surfaces.

In Figure 3(b) we show similar results from averaging over the first 100 ps of the simulation for iBu. By comparing Figure 3(b) with 3(a), we can see that some molecules moved up one layer (such as molecule-3, 16, and 22) and some moved down one layer (such as molecule-5, 20, 23, and 27) between the first 100 ps and the second 100 ps of dynamics. We also can see that the spread of some molecules has changed. For example, the spread of molecule-15, 17, and 25 was less, which implies that the orientation of these molecules shifted more toward the x,y-plane so that a smaller length is projected along the z-axis. This is consistent with the earlier finding that Lennard-Jones molecular chains reorient themselves to the shear direction in ultra-thin films *(6)*. On the other hand, the spread of molecule-9, 10, 14, and 18 was increased. Apparently, these molecules were oriented more toward the z-axis under the shear. Notice that these molecules are in the intermediate layers of the lubricant, which are absent in ultra thin films. We think that it is possible for the molecules in intermediate layers of a "medium-thin" film such as the one simulated here to reorient against the shear direction. The molecules at the top and bottom layers in our simulation yield well-defined narrow distribution bands in both Figure 3(a) and (b) and are oriented well in the x, y-plane that contains the shear direction.

Stick-Slip Motions

Figure 4 shows the displacement of the center-of-mass of each lubricant molecule, along the shear direction as a function of time. In the figures the diagonal black line and the horizontal black line represent the displacements of the top and bottom SAM molecules, which clearly indicate a velocity of 1 Å/ps for the top SAM and a zero velocity for the bottom SAM. To distinguish the motion of each molecule, we partitioned the 32 lubricant molecules into 8 layers according to the similarity of their motion in the x direction. Each layer is indicated with a distinct color. The label for each molecule in the figure for each case is consistent with the numbering of the molecule shown in Figure 3 (and also in Figures 5, 6, and 7).

We can see *stick-slip* motion in the molecules near the bottom interface in all three cases, characterized by a sequence of move (slip), pause (stick), and move (slip) again. The case of iBu shows profound stick-slip character, with a periodicity of $\sim$40 ps. The stick-slip motion in the molecules near the top interface for iBu is also visible (here the stuck state refers to the top SAM layer). For iPr and Ph, the molecules near the top interface show constant sliding after 100 ps.

The stick-slip motion is often observed in macroscopic studies on solid surfaces *(4,15)*. When a solid block, placed on top of a solid surface, is driven elastically, the stick-slip motion is observed. The stick-slip phenomena is also observed in ultra-thin films *(5, 16, 17)*. Here, we observed the stick-slip motion in lubricant films of "medium-thin" thickness. In our simulation, the top "brush" surface was driven smoothly at a constant velocity, and we observed the stick-slip motion in both bottom and top layers of the lubricant. In this case the stick-slip motion of the interfacial boundary propagates into the lubricant.

What is the origin of the lubricant stick-slip motion? In the case of stick-slip sliding on solid surfaces and in ultra-thin films, there is a one-to-one correspondence between the stick-slip motion and the frictional force or shear stress (the ratio of the frictional force to the contact area). The friction is highest during the stick period (static friction) and lowest during the slip period (kinetic or dynamic friction). We have evaluated the frictional forces in our systems, but unfortunately, the fluctuations of the frictional forces themselves are so large that it is impossible to differentiate them from the difference between static friction and kinetic friction *(11)*. However, the fluctuations of frictional forces in our system are too large to be explained in terms of pure thermal fluctuations. We think that the frictional forces are responsible for the observed stick-slip motion in the lubricant.

Since the stick-slip behavior of the lubricant molecules changes with the nature of the wear inhibitor molecules at the interface, we concluded that the properties of the SAM molecules must play a role. Since the SAM molecules stay at the same relative position on the iron oxide surface, they probably modulate (elastically)

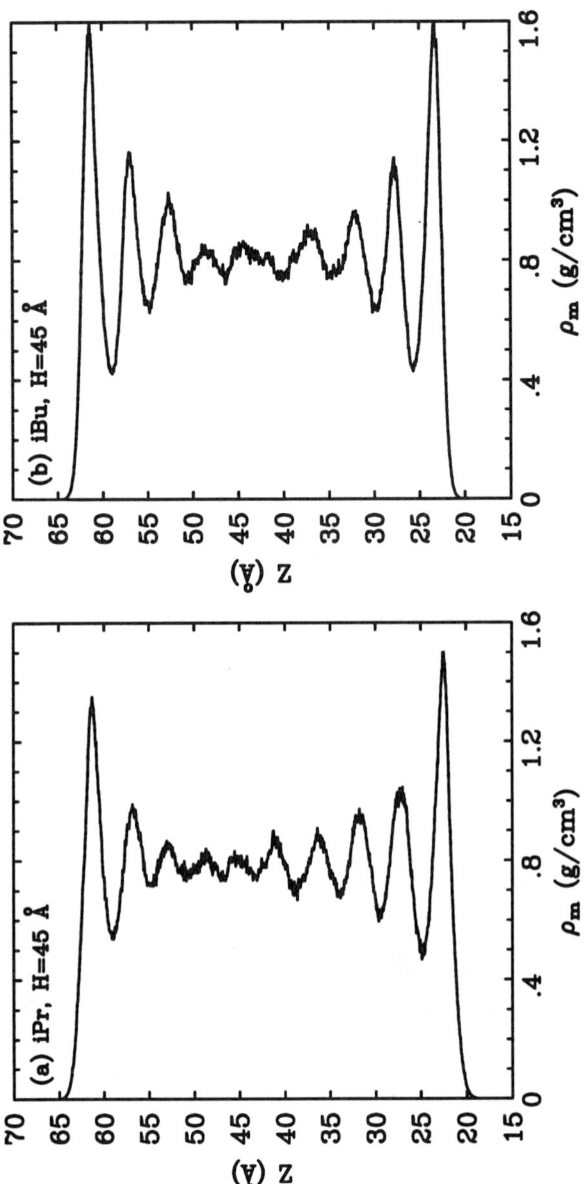

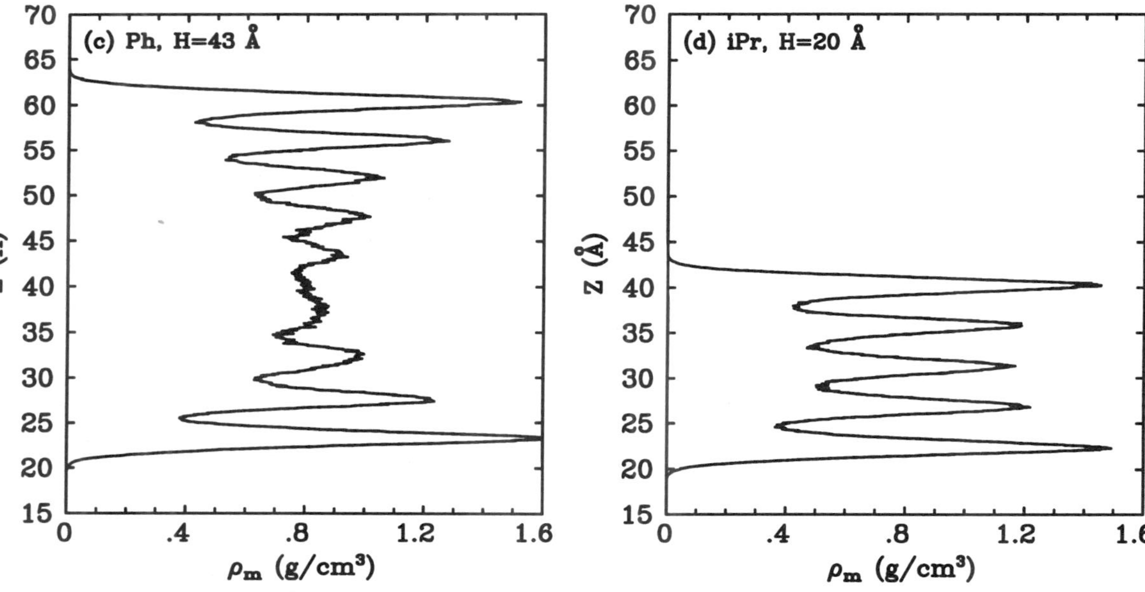

Figure 2. The density distribution of the lubricant as a function of z (perpendicular to the surface). We used 1000 bins to count the number of atoms falling in a zone between z and z + dz at each time-step. This density was averaged over the last 100 ps of the 200 ps runs: (a) iPr, H~45 Å; (b) iBu, H~45 Å; (c) Ph, H~43 Å; and (d) iPr, H~20 Å.

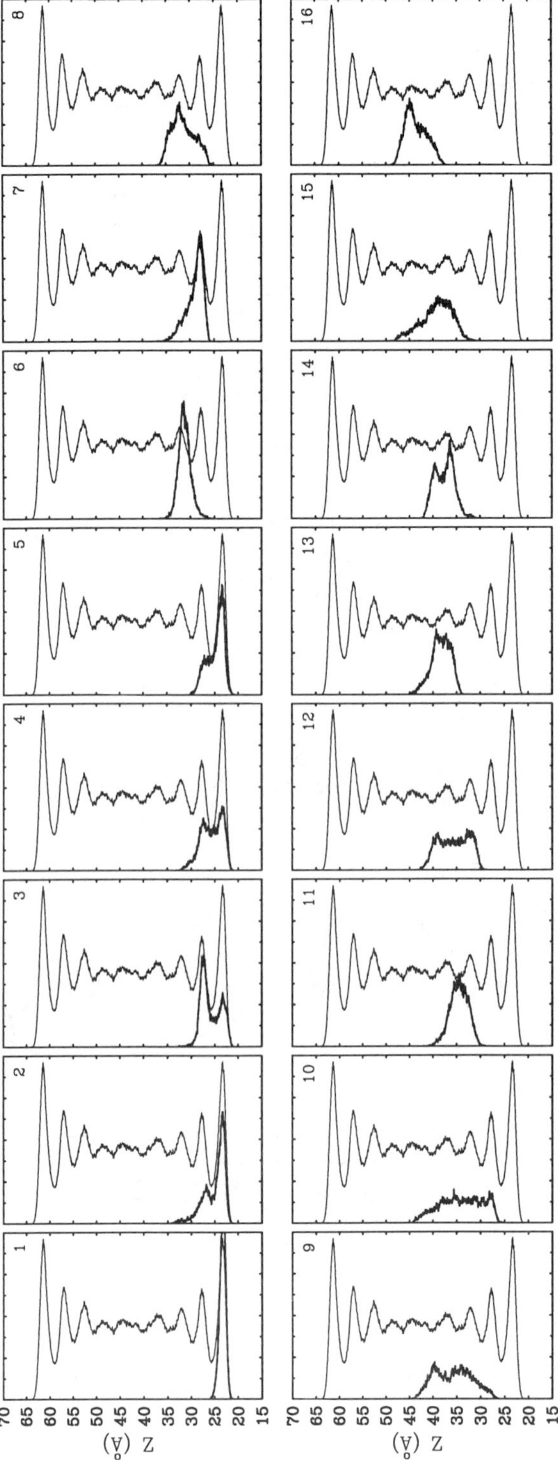

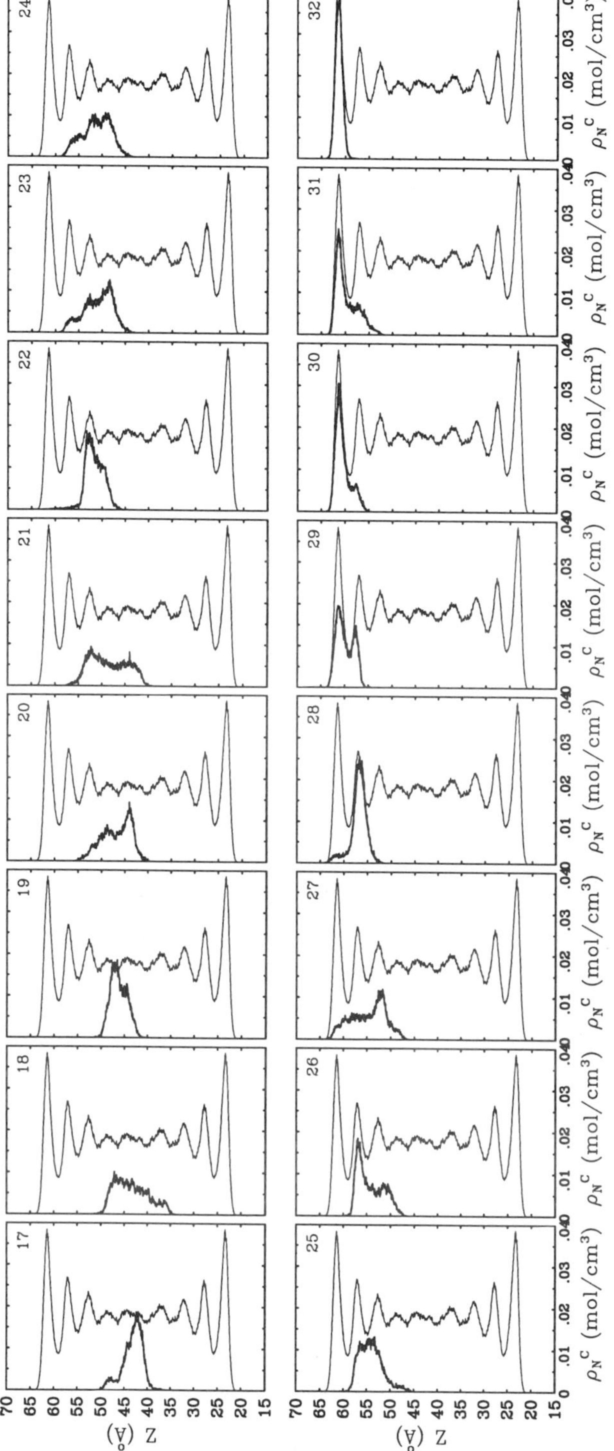

Figure 3. The distribution of carbon atoms for each lubricant molecule (dark solid curve) for the iBu case. The background (light solid) curve is the distribution of total carbon atoms of all molecules. (a) is the results of averaging over last 100 ps of the dynamics, and (b) the results of averaging over first 100 ps of the dynamics. Continued on next page.

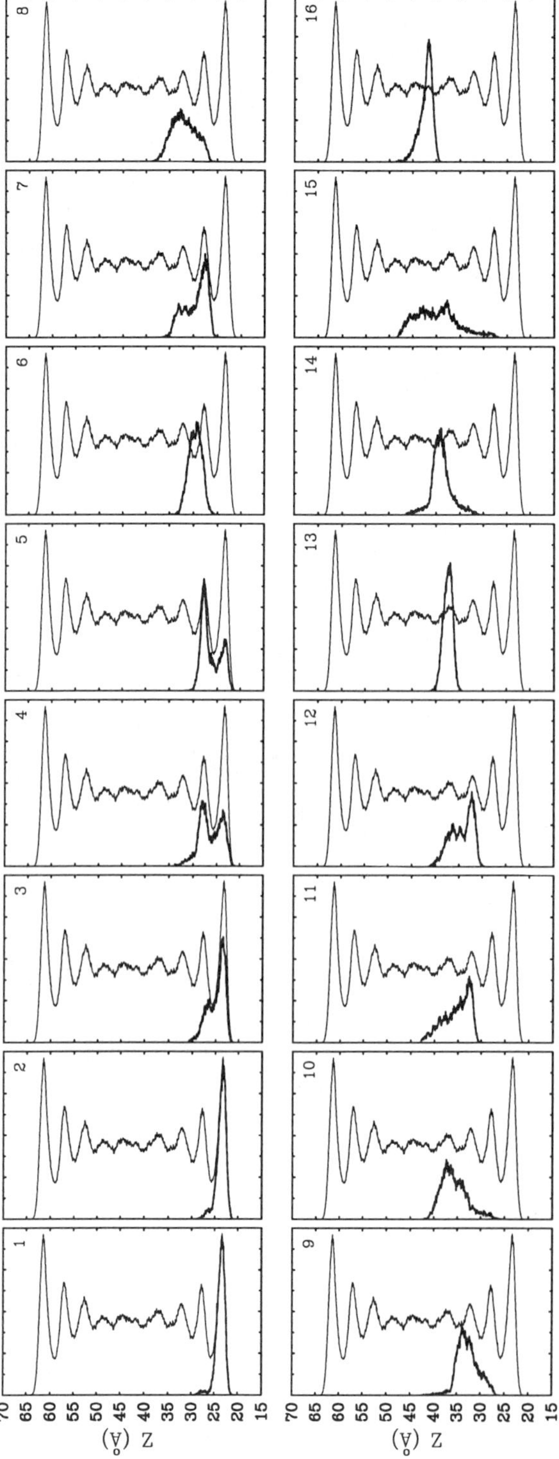

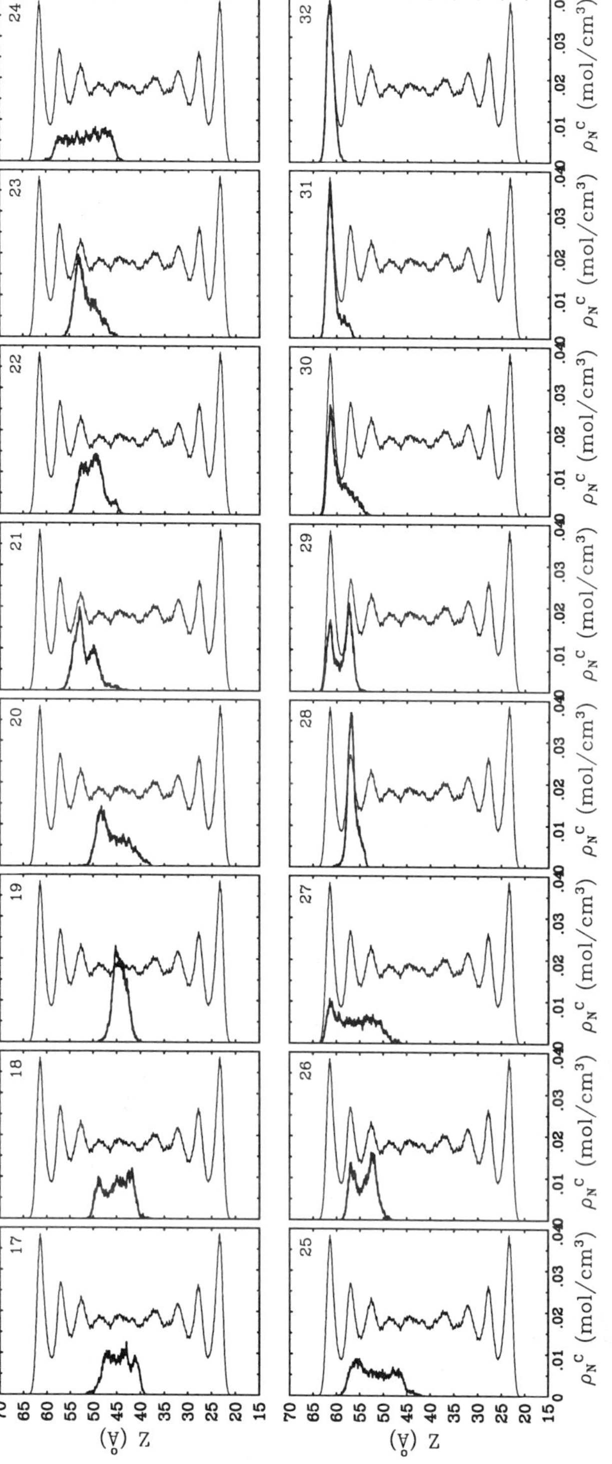

Figure 3. Continued.

the frictional forces on the lubricant. Thus, although the top wall is moving at a constant velocity, the wear inhibitor molecules lead to an elastic response causing oscillatory driving forces on the lubricant. That is, the lubricant might compress or push the SAM molecules which then rebound to press on the neighboring lubricant molecules. A detailed study of this issue will be the subject of future work.

For ultra-thin films, Boljon and Robbins *(6)* found that change from slip to stick is associated with a phase transition from melt or glassy state to solid of the film. In their case the liquid film is so thin that the motion of the entire film can be collectively characterized by either stick or slip. However, in our simulation, the lubricant molecules in each layer are in the stick (or slip) state at different times. For instance, in the case of iBu, molecule-2 at the bottom is in stick from 120-160 ps, whereas molecule-31 at the top are in stick from 95-135 ps. Because of this inhomogeneity, we do not observe a phase transition in our lubricant film of the corresponding to the change between stick and slip.

Structures at Various Stick States

To examine the structure of the lubricant molecules when they are in stick, and provide a better microscopic picture of the lubricant molecules, we provided a series snapshots of two molecules selected from the case of iBu: one from the bottom and one from top.

Figure 5(a) shows the top and side views of molecule-2 (near the bottom) between 120-165 ps. Only SAM molecules are shown in the background as a reference. The side views show that this molecule stands up a little bit on the top of the SAM layer (namely surface). Only one end-segment (low-left corner) of this molecule lies on the top of the SAM layer. The molecule was in stick during 120-160 ps. At 165 ps, when the whole molecule is sliding, the portion of this end-segment is smallest, compared with that at earlier time instances. From the top views, it appears that this end-segment falls behind the other portions of the molecule and that it is the far end-segment of the molecule (away from the surface) that is driven ahead by the shear flow. During 120-160 ps while the molecule struggles from sliding, it moves back and forth in the y-direction.

Figure 5(b) show the snapshots of a top molecule, molecule-31, during 95-140 ps. Since the stick-slip motion of the top molecules refers to the top wall, in plotting this molecule we translated the top SAM layer at different times back to its original position according to the corresponding shear distances. Therefore, the top lubricant molecules in effect experience a shear flow in the $(-x)$-direction.

Side views in Figure 5(b) show that molecule-31 more or less lies on the x, y-plane at 95 ps, but later reoriented itself towards the z-axis (more spread along the z-axis). During 95-115 ps, when the molecule is in stick with the top SAM layer, both end-segments of the molecule are dragged by the negative shear flow ("bottom" views). At 120 ps, the molecule has a small slip. Between 125 ps to 135 ps the molecule gets stuck again. At 140 ps, the molecule begins sliding. Again, both of its end-segments are dragged by the negative shear flow and moving ahead of the central portion of the molecule.

These snapshots indicate the conformational complexity of the lubricant molecules during the stick-slip motion. There are always certain portions of a molecule being dragged by the shear flow and moving ahead than other portions of the molecule. Sometimes it is the end-segments, and in other times it is the central portion of the molecule. Which portion of the molecule becomes the dragging front depends on the amount of interactions between the portion of the molecule and the nearby SAM layer.

Radius of Gyration, End-To-End Distance, and Torsional Conformation Distribution

We calculated the radius of gyration (R_0), end-to-end distance (L), and number of *trans*-conformers (N_t) for each lubricant molecule. For the sake of brevity, we show only the results for ten molecules from the iBu case (five from bottom, five from top). In Figure 6, we see strong correlations between these properties in all three cases. When the molecule is stretched, the radius of gyration increases, so do the end-to-end distance and the number of *trans*-conformers in the carbon chains.

Figure 6 also clearly indicates that there is no correlation between these properties and the stick-slip motion of the molecules. Apparently, when the molecule gets stuck, it is equally possible for it to be stretched (such as molecule-2 from 120 ps to 125 ps) or recoiled (such as molecule-31 from 110 ps to 115 ps). Changing from slip to stick state can either increase or decrease the end-to-end distance of the molecule. In addition, in situations where the molecule appears in the stick state in the shearing direction but it actually moves in one of the other two perpendicular directions, information on the motion in the shearing direction alone (which defines the stick-slip motion) cannot determine whether the molecule has been stretched or compressed. Thus, we did not observe a one-to-one relationship between the stick-slip motion and the variation of the radius of gyration or the end-to-end distance of the molecules.

In Figure 7, we plotted the radius of gyration, end-to-end distance, and number of *trans*-conformers for each molecule averaged over the last 100 ps of the simulations against the average z-position of the center-of-mass of the molecule. Here, besides the fluctuations, there are no systematic changes in these quantities across the lubricant films (*i.e.*, along the z-axis). The results indicate that stretching the top and bottom molecules are equally likely even though the top molecules have much higher shear velocity than the bottom ones, which is consistent with Gallilean invariance.

Conclusion

With the advances of computer simulation techniques we are able to simulate rather realistic and complex systems. Here we simulated nanoscale hexadecane

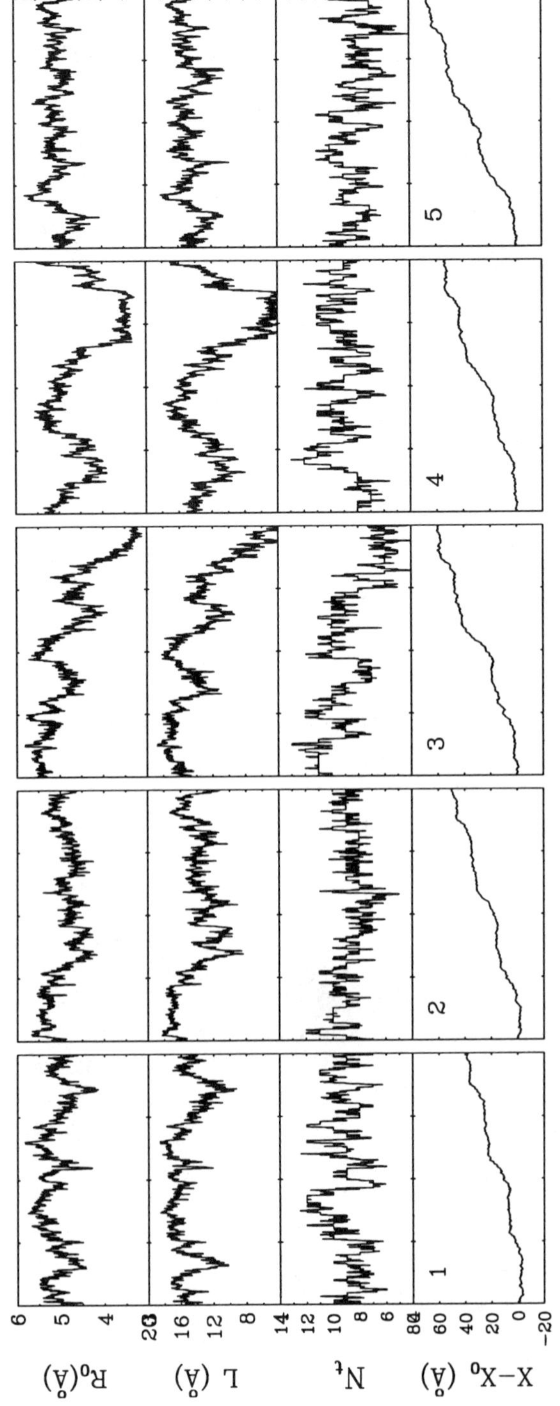

R$_0$(Å)
L(Å)
N$_t$
X–X$_0$(Å)
6 5 4
20 16 12 8
14 12 10 8 6
80 60 40 20 0 –20
1 2 3 4 5

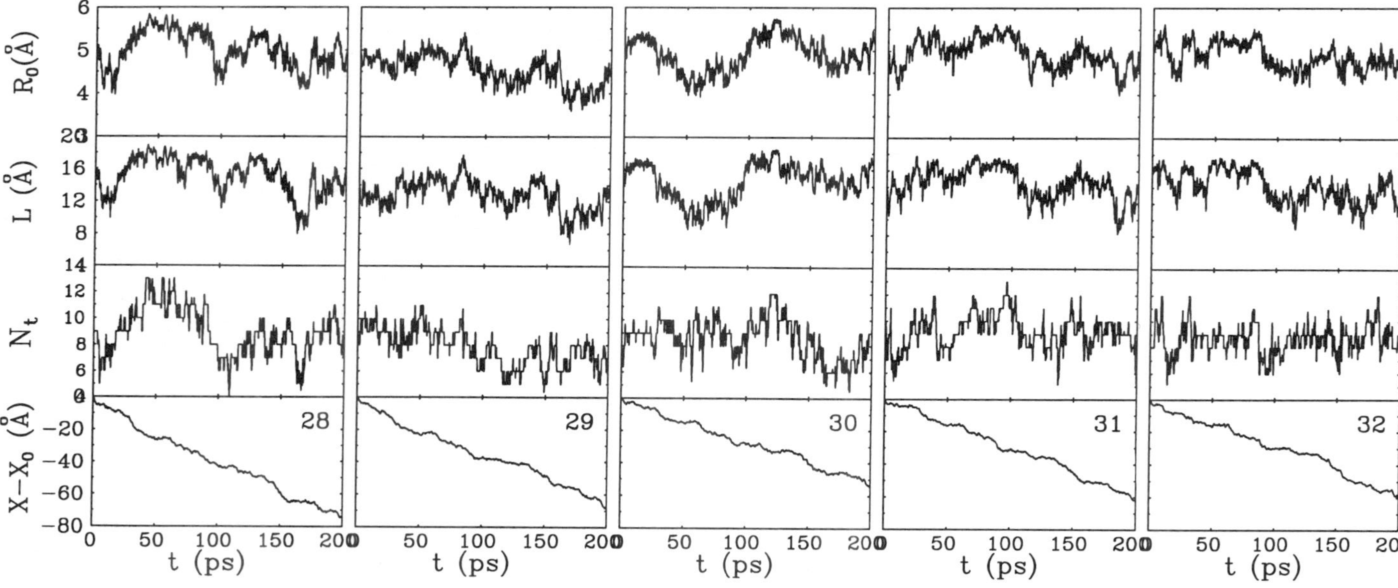

Figure 6. The radius of gyration (top), end-to-end distance (middle), number of trans-conformers (bottom), and displacement in the x-direction as a function of time, for the lubricant molecules in the case of iBu. (a) Molecule-1, 2, 3, 4, and 5 from the bottom; and (b) molecule-28, 29, 30, 31, and 32 from the top. The numbering of these molecules is the same as in Figure 3.

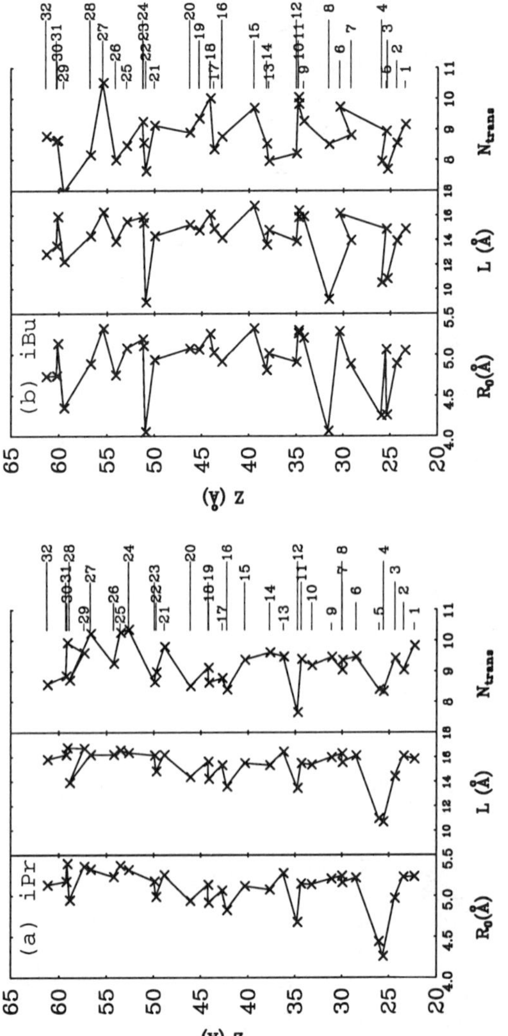

(b) iBu
(a) iPr
Z (Å)
N_{trans}
L (Å)
R_0(Å)

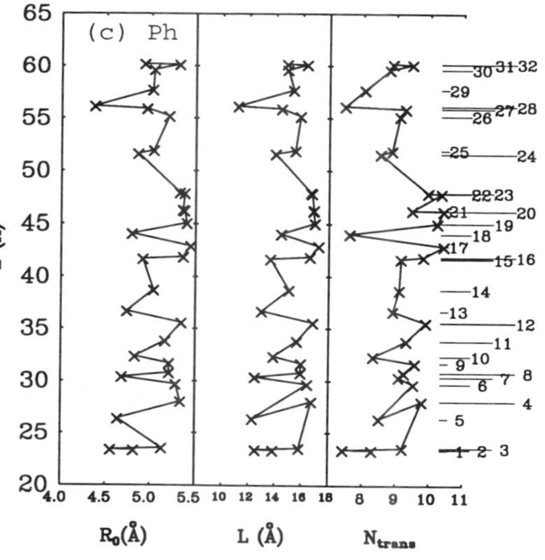

Figure 7. *The average radius of gyration (left), end-to-end distance (center), and and number of trans-conformers (right) of each lubricant molecule as a function of its average position in z. (a) For iPr, (b) for iBu, and (c) for Ph.*

lubricant films confined between two "brush" surfaces consisting of iron-oxide surface and adsorbed wear inhibitor molecules. We found that in such a confined geometry, the lubricant films exhibit strong density oscillations. For the separations of $\sim$44 Å we found 9-10 layers in the density, and for a $\sim$20 Å separation we found 5 layers. In addition, we found that the motion of each individual molecule in the direction perpendicular to the surfaces is very limited. They are confined to within 1-2 layers.

Under a shear flow, the lubricant molecules near both bottom and top surface boundaries show stick-slip motion in the shear direction. The stick-slip motion cannot be simply characterized by the stretching or compressing of molecules. The characteristics of stick-slip motion are very sensitive to the type of "brush" molecules on the surfaces that interact directly with the lubricant molecules.

Acknowledgments—This research was supported by the Chevron Chemical Company (Oronite Technology group). The development of the methodologies used here was supported by grants from the DOE-ASCI-ASAP and NSF (CHE 95-12279 and ASC 92-17368). The facilities of the MSC used in this research are also supported by grants from ARO-DURIP, BP Amoco, ARO-MURI (Kiserow), Beckman Institute, Seiko-Epson, Exxon, Avery-Dennison Corp., 3M, Dow, and Chevron Research Technology Co.

References

1. Bhushan, B.; Israelachvili, J. N.; Landman, U. *Nature* **1995**, *374*, 607.
2. Thompson, P. A.; Grest, G. S.; Robbins, M. O. *Phys. Rev. Lett.* **1992**, *68*, 3448.
3. Granick, S. *Science* **1991**, *253*, 1374.
4. Yoshizawa, H.; Israelachvili, J. N. *J. Phys. Chem.* **1993**, *97*, 11300.
5. Lupkowski, M.; Swol, F. c. *J. Chem. Phys.* **1991**, *95*, 1995.
6. Baljon, A. R.; Robbins, M. O. *Micro/Nanotribology and its Applications;* Bhushan, B. Ed.; Kluwer Academic Pub.; Amsterdam, 1997; pp 533-553.
7. Georges, J. M.; Tonick, A.; Poletti; S.; Yamaguchi, E. S.; Ryason, P. R. *Trib. Trans.* **1998**, *41*, 543.
8. Jiang, S.; Dasgupta, S.; Blanco, M.; Frazier, R.; Yamaguchi, E. S.; Tang, Y.; Goddard, W. A., III *J. Phys. Chem.* **1996**, *100*, 15760.
9. Jiang, S.; Frazier, R.; Yamaguchi, E. S.; Blanco, M.; Dasgupta, S.; Zhou, Y.; Çağin, T.; Tang, Y.; and Goddard, W. A., III *J. Phys. Chem. B* **1997**, *101*, 7702.
10. Zhou, Y.; Jiang, S.; Çağin, T.; Yamaguchi, E. S.; Frazier, R.; Ho, A.; Tang,Y.; Goddard, W. A., III *J. Phys. Chem. A* **2000**, *104*, 2508.
11. Zhou, Y.; Çağin, T.; Yamaguchi, E. S.; Frazier, R.; Ho, A.; Tang, Y.; Goddard, W. A., III *J. Phys. Chem.* **2000**, submitted.
12. Mayo, S. L.; Olafson, B. D.; Goddard, W. A., III *J. Phys. Chem.* **1990**, *94*, 8897.

13. Rhykerd, C. L. Jr.; Schoen, M.; Diestler, D. J.; Cushman, J. H. *Nature 1987, 330*, 461.

14. Stevens, M. J.; Mondello, M.; Grest, G. S.; Cui, S. T.; Cochran, H. D.; Cummings, P. T. *J. Chem. Phys. 1997, 106*, 7303.

15. Yoshizawa, H.; Chen, Y. L; Israelachvili, J. N. *J. Phys. Chem. 1993, 97*, 4128.

16. Thompson, P. A.; Robbins, M. O. *Science 1990, 250*, 792.

17. Robbins, M. O.; Baljon, A. R. *Microstrcture and Microtribology of Polymer Surfaces;* Tsukruk, V. V.; Wahl, K. J. Eds.; American Chemical Society Symposium Series; Washington, DC, 1999; Vol. 741, pp 91-115.

Chapter 13

Complete Spreading of Liquid Droplets on Heterogeneous Substrates: Simulations and Experiments

M. Voué[1], S. Semal[1], C. Bauthier[1], J. J. Vanden Eynde[2], and J. De Coninck[1]

[1]Centre de Recherche en Modélisation Moléculaire and [2]Laboratoire de Chimie Organique, Université de Mons-Hainaut, 20, Place du Parc, B–7000 Mons, Belgium

We report the results of molecular dynamics (MD) simulations probing the spreading of a chain-like liquid droplet on heterogeneous substrates. A comparison is carried out between patterned and random substrates made of A and B materials. The simulations show that the spreading rate of the liquid droplet is a monotonous function of the heterogeneity concentration. The spreading of low molecular weight silicon oils (PDMS) on heterogeneous substrates has also been experimentally studied using high resolution spectroscopic ellipsometry. Two types of substrates were investigated : partially OTS-grafted silicon wafers and silicon wafers whose coating was a two-component grafted layer (UTS and uUTS). In the former case, the reaction time was varied to change the heterogeneity concentration at the wafer surface, while in the later, the relative concentrations of the coating constituents in the grafting solution were modified. The experimental results are in close agreement with the theoretical predictions of the MD simulations.

Introduction

Numerous theoretical descriptions of the "solid-liquid" interaction are based on *ab initio* methods or on first principles computations. These computational methods are used in a view of investigating the bulk or the interfacial properties of liquids molecules in contact with a solid surface (*see the others contributions in this volume*). To that purpose, an increasing - but still restricted – number of atoms is considered and the related force-fields take into account, as much as possible, the specificity of the "solid-liquid" interaction and its chemical complexity. As in most of the fields of interest related to the microscopic aspects of the supramolecular aggregation or of the collective motion of molecules, the theoretical description of the spreading phenomenon (i.e. the phenomenon according to which a liquid droplet interacts with a solid surface but also with its vapor phase) requires to obtain a compromise between two factors : the size of the system and the chemical complexity of the interactions. The computational complexity of the problem is mainly related to the displacement of the contact line, at which the three phases meet each other. But, to some extent, the size of the system seems to be the relevant factor to be considered in order to recover in the computer experiments the basic features of the physical phenomenon, such as the contact line dynamics (for a review, see e.g. Voué, M. ; De Coninck, J. *Acta Materialia* in press). Recent attempts have been made to combine the chemical complexity with the wetting phenomenon using the ES+ force field (Swiler, T.P. ; Loehman, R.E., J. *Acta Materialia* in press) but this kind of approach still suffers from a lack in considering a large number of atoms in the simulation. To overcome this inconvenience, in our computational approach of wettability, we consider large number of atoms for both the solid (typically 300,000) and the liquid (typically 6,000 to 25,000) and simplified force fields (i.e. short range interactions as described by the standard 6-12 Lennard-Jones potential), as described in Section II of this contribution. This kind of approach should be considered as a medium to coarse grain simulations.

More specifically, wetting properties of heterogeneous surfaces are important to understand because they influence strongly either the static or the dynamics properties of the liquid that come into contact with them (*1*). The dynamics of the contact line is one of the most studied phenomenon resulting from the presence of these heterogeneities and some numerical model, mainly based on the Monte-Carlo theories have been proposed during the past 20 years (*2*). Recently, molecular dynamics (MD) simulations were performed to investigate, in the partial wetting regime, the time relaxation of the contact angle of a liquid droplet on an heterogeneous substrate (*3, 4*). The authors of that study concluded that the wetting dynamics is adequately characterized by the molecular-kinetic theory of wetting (*5-6*) on random substrates and on substrates with heterogeneity patches.

In this contribution, we address the same problem but in the complete wetting regime to see whether this equivalence still exists or has to be reconsidered. The structure of the article is the following : the characteristics of the complete spreading at the microscopic scale are reviewed in Section I. MD simulations results are presented in Section II while the results of experimental investigations are reported in Section III. The conclusions of our study are presented in Section IV.

Complete spreading at the microscopic scale

Experimental approach.

Since the experimental work of Heslot (*7-9*) and theoretical studies of de Gennes and coworkers (*10*) on the stratified droplets, the pseudo-diffusive behavior of the liquid molecules on homogeneous substrates is well established. Experimentally, the length of the precursor film (of molecular thickness) which grows in front of the macroscopic part of the droplet scales as

$$l_{film} = \sqrt{Dt} \tag{1}$$

allowing to define a pseudo-diffusion coefficient D for the liquid molecules on top of the solid substrate. The spreading phenomenon is now clearly understood as a competition between the affinity of the liquid molecules for the solid substrates, which acts as a driving term, and the friction at the microscopic scale (*11*) :

$$D = \frac{\Delta W}{\zeta} \tag{2}$$

where ΔW is here the difference of energies of the molecules in the precursor film and in the reservoir and ζ is the friction coefficient of the liquid molecules on the solid substrate.

Computational approach.

A few years ago, using Molecular Simulations (MD) techniques, De Coninck and coworkers could recover both the terraced spreading phenomenon and the $t^{1/2}$ law (*12-13*) but these numerical results were obtained using a large number of liquid atoms in such a way that the central part of the drop acts as a reservoir, an atomic description of the solid and an appropriate temperature control of the solid substrate. These authors have considered 25,000 atoms for the fluid and 250,000 atoms for the solid. The microscopic details of the mechanism of spreading and, in particular, the role of the friction at the microscopic level, were recently also investigated using MD simulations (*11*). The results of these studies are in agreement with experimental evidences and support the validity of the de Gennes-Cazabat model based on driving force/friction competition mechanism (*10*).

Disordered substrates.

This computational approach has been recently generalized to heterogeneous substrates and experimentally validated (*14*). An analytical expression has been given to express D_{AB} the diffusion coefficient of the liquid molecules on an heterogeneous substrate made of A and B materials at relative concentrations C_A and C_B as a function of D_A and D_B, the diffusion coefficients measured on the corresponding pure substrates :

$$D_{AB} = \frac{C_A D_A + C_B D_B \Lambda}{C_A + C_B \Lambda} \tag{3}$$

where $C_B = 1 - C_A$ and where $\Lambda = \zeta_A / \zeta_B$ is the ratio of the friction coefficients of the liquid molecules on the pure substrates.

Nevertheless, this equation can only be considered as a first approximation because it only relates the diffusion coefficient of the liquid molecules on the heterogeneous substrates to the heterogeneity concentration, neglecting the effect of the size and of the geometry of the heterogeneity patches. Some of these effects are considered hereafter.

Molecular Dynamics

Models.

As in previous studies (*12-14*), all the particles considered in this study interact with each other via a standard 6-12 Lennard-Jones potential

$$U_{ij}(r) = 4\varepsilon \left[\left(\frac{\sigma}{r} \right)^6 - \left(\frac{\sigma}{r} \right)^{12} \right] = \frac{C_{ij}}{r^6} - \frac{D_{ij}}{r^{12}} \tag{4}$$

In the equation 4, ε is the depth of the potential well, σ is the characteristic radius of the particles and r denotes the distance between a pair of particles i and j. Only short-range interactions are taken into account : the Lennard-Jones interactions are cut-off for $r \geq 2.5\sigma$. The C_{ij} and D_{ij} parameters are chosen constant, for simplicity. For the fluid-fluid (ff) interactions, their value is $C_{ff} = D_{ff} = 1.0$.

To describe the behavior of chainlike molecules, neighboring atoms of a given molecule interact via an intramolecular confining potential which scales, for computational convenience, as the 6th power of the distance separating the atoms :

$$U_{int}(r) = r^6 \tag{5}$$

In this study, we considered 6400 particles grouped into 400 16-atoms chains to mimic the behavior of the liquid. As shown in (*12-13*), the size of this system, although smaller than the one usually considered, is large enough to recover the $t^{1/2}$ spreading law.

The solid substrates on top of which the liquid droplet will spread are composed by two layers of particles (about 45,000 particles). These solid particles interact with each other via a Lennard-Jones potential, as given by equation 4, with parameters C_{ss} = 35.0 and D_{ss} = 5.0. They are heavier than the liquid atoms ($m_{solid} = 50\, m_{liquid}$) so that their displacement has a time scale comparable with the liquid. These solid particles are initially placed on a (100) f.c.c. lattice and their motion around these initial positions is restricted with a harmonic potential.

The substrate is composed of two materials A and B which will interact differently with the liquid molecules. Let us suppose that B is more wettable than A. For this reasons, we have chosen to use the following sets of parameters for the fluid-solid (fs) interactions : $C_{fs}^{A} = D_{fs}^{A} = 0.8$ and $C_{fs}^{B} = D_{fs}^{B} = 2.0$. Although differences in wettability exist between A and B, both kinds of fluid-solid interactions are relevant of the complete wetting regime (*15*) : a liquid monolayer completely covers the substrate and the equilibrium contact angle is apparently equal to 0. The complete spreading of the liquid molecules (with interaction parameters $C_{ll} = D_{lls} = 1.0$) on the solid substrate with interaction $C_{fs}^{A} = D_{fs}^{A} = 0.8$ appears to be a direct consequence of the mass difference between the atoms of the liquid and of the solid.

In order to compare the influence of the substrate geometry on the complete spreading of the liquid, we have considered in the present study two kinds of substrates : (i) random substrates and (ii) substrates of type A with regular rectangular patches of B atoms. For the latter case, the influence of the heterogeneity concentration has been investigated by increasing the size of the patches.

Given the potentials, the motion follows by integrating Newton's equation of motion using a fifth order predictor-corrector method. The time step Δt is measured in units of the dimensionless time variable $\tau = (\varepsilon m)^{1/2}/\sigma$ and its value given by $\Delta t = 0.005\, \tau$ is of the order of 5 10^{-15} s, with ε and σ defined as above.

Although all the sets of simulations reported in our previous publications (*12, 13, 15*) are consistent with each other, the choice of the parameters is – to some extent – crucial for the comparison with the experiments. To compare the simulations results with the dry spreading experiments carried out with the silicon oil (see Section III), one has to mimic the non-volatility of the liquid and absence of constraint in the bending angle of the chains. This is easily achieved using the force field that we have chosen. Examples concerning the parameterization of the force field and their consequences on the simulation results are given in (*26*).

The temperature of the system is controlled by rescaling the velocities of all the particles. The liquid droplet and the substrate are equilibrated independently from each other. About 50 10^3 time steps later, i.e. as soon as the droplet energy reach a constant value, it is moved into the vicinity of the substrate and starts to spread. From that time, only the temperature of the substrate is controlled, allowing heat exchange between the liquid and the solid, to mimic energy dissipation as it occurs in a real experimental setup.

Simulation results.

On the basis of our previous studies (*12, 13, 15*), one may expect that the larger the concentration of B is, the faster the spreading should be. Starting from a pure substrate characterized by coupling constants $C_{fs} = D_{fs} = 0.8$ (material A), the heterogeneities are build up as squares of atoms with a coupling $C_{fs} = D_{fs} = 2.0$ (material B). Either by varying the size of these more wettable patches of B or by increasing their number at random positions, we investigated heterogeneity concentrations equal to 0.0, 8.2, 44.4, 73.4 and 100.0 percent.

Typical snapshots and typical density profiles are given in Figure 1. It can be seen that on the patterned substrates (Figure 1b), three regions can be identified in the liquid drop : (a) the reservoir, (b) an intermediate compact liquid layer and (c) an outer region whose structure is highly dependent on the substrate structure. In that region, the liquid molecules cover only the atoms of the substrate characterized by $C_{fs} = D_{fs} = 2.0$, although, on the basis of what is seen on the pure substrates, they completely wet both types of substrate atoms. This effect is not - or less - visible in the intermediate region because the liquid molecules have more neighbors and that the liquid-liquid interactions compensate the trends for these molecules to partially dewet the substrate. On the random substrates (Figure 1b), the contact line is less perturbed by the heterogeneities than on the patterned ones. This point will be discussed in details in a forthcoming article.

As the spreading rate is determined from the growth of the precursor film, we calculated the radius of the film - assuming a radial symmetry - from the sigmoid shape of the density profile of the layer (Figure 2a). As the corresponding base radius R_1 is scales as $t^{1/2}$ (Figure 2b), we can therefore calculate the diffusion coefficient of the liquid molecules on the substrate. This was done for increasing concentrations of heterogeneities either on the patterned or on the random substrates. The results are presented in Figure 3. D decreases monotonously with the concentration C_A of the

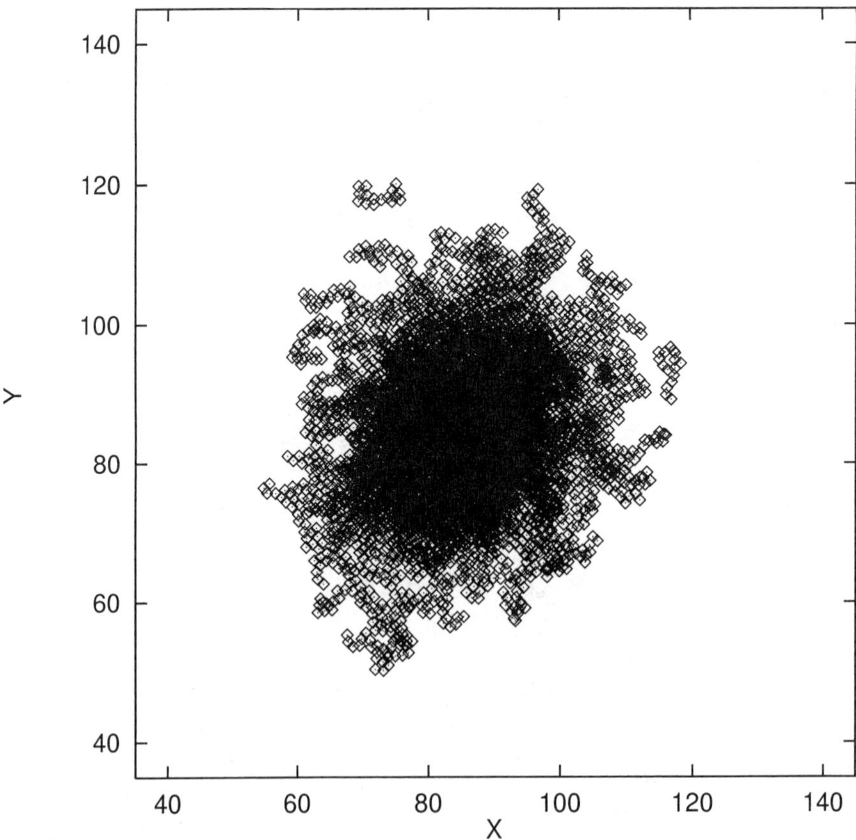

Figure 1. Snapshots of a droplet spreading on homogeneous and heterogeneous substrates (top views) after 100 10³ time steps. Only the heterogeneities present at the solid substrate are represented by dots. (a) Homogeneous substrate : $C_{fs} = D_{fs} = 0.8$; (b) Random substrate : 44.4 % of the solid atoms interact with the fluid molecules with coupling constants $C_{fs} = D_{fs} = 2.0$; (c) Patterned substrate : same concentration and parameters as in (b) ; (d) Homogeneous substrate : $C_{fs} = D_{fs} = 2.0$.

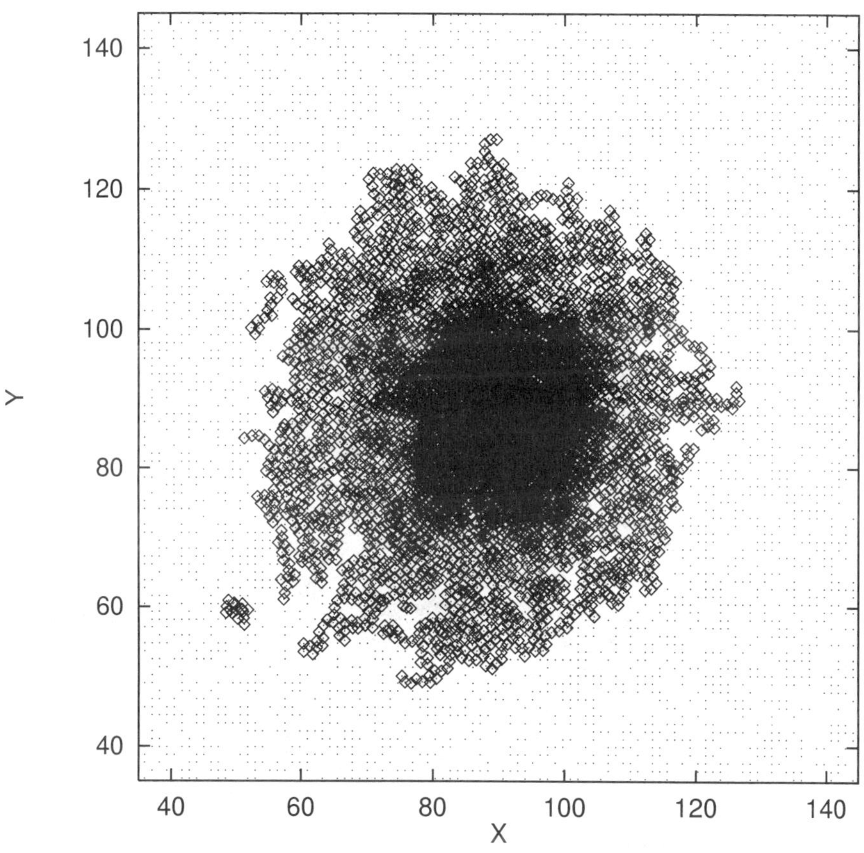

Figure 1. *Continued.*

Continued on next page.

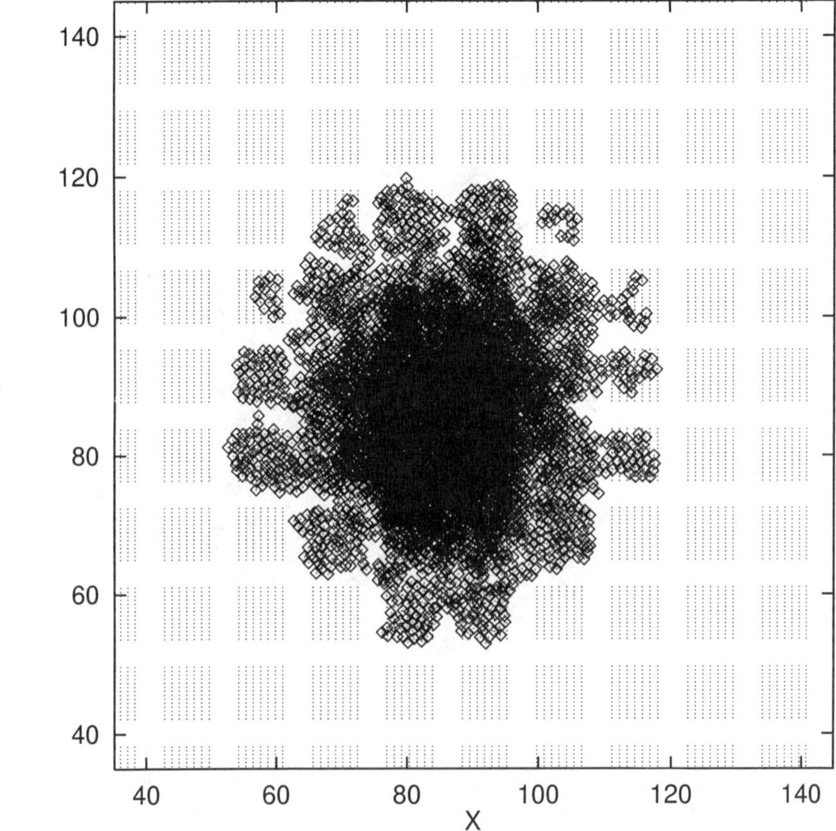

Figure 1. *Continued.*

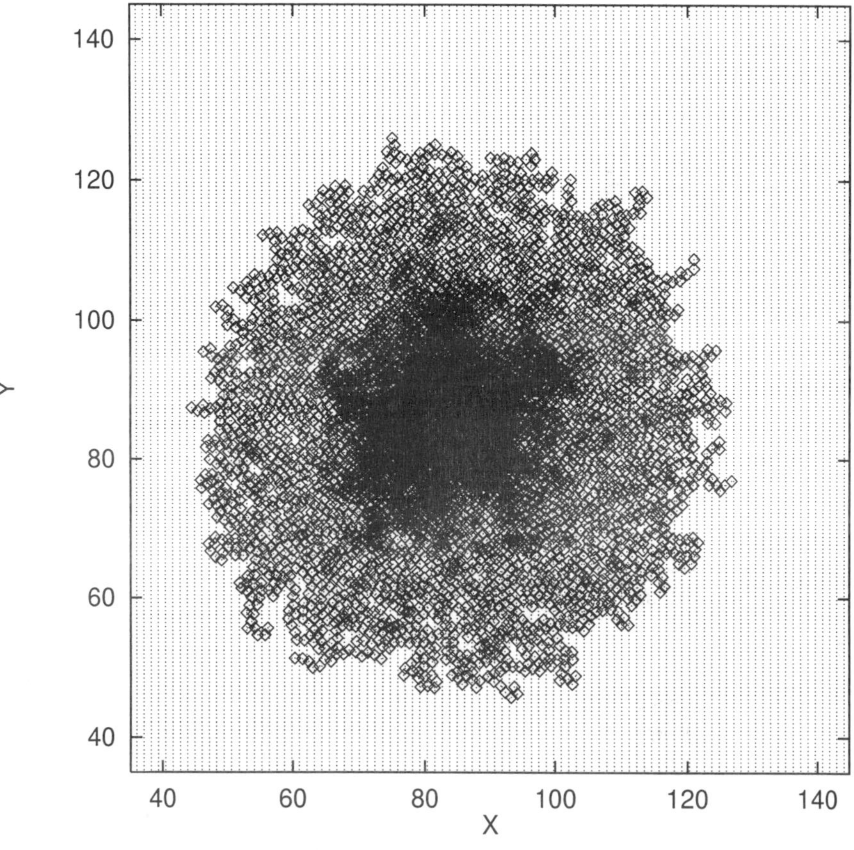

Figure 1. *Continued.*

188

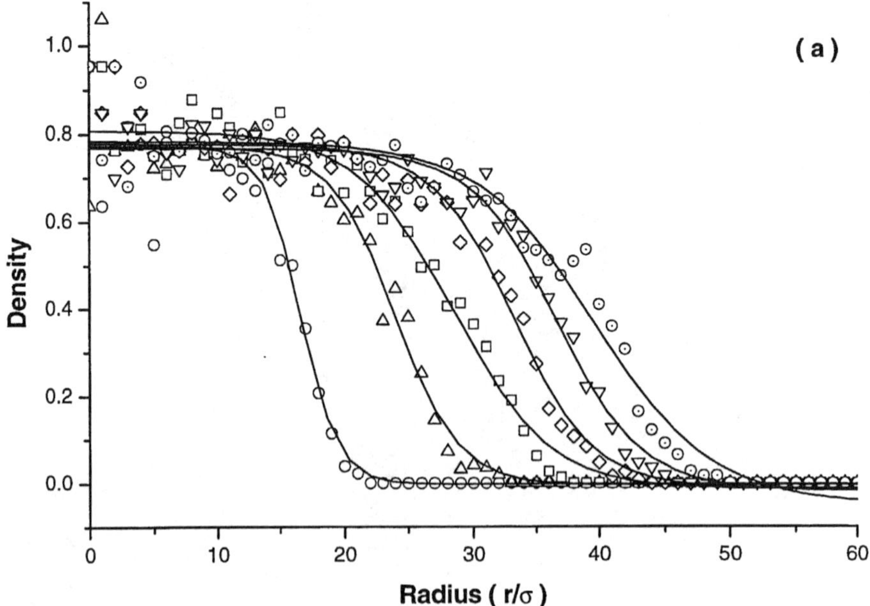

Figure 2. (a) Density profiles of the liquid molecules spreading on a random
substrate as a function of the distance from the droplet *symmetry axis. The profile has*
have been calculated at increasing time (from left to right : 27.0, 32.0, 37.0, 42.0,
47.0 and 52.0 10^3 time steps). 44.4 % of the solid atoms are characterized by solid-
liquid coupling constants $C_{fs} = D_{fs} = 2.0$ *; (b) Corresponding time evolution of the*
precursor film contact area with the solid. It linearly grows with t.

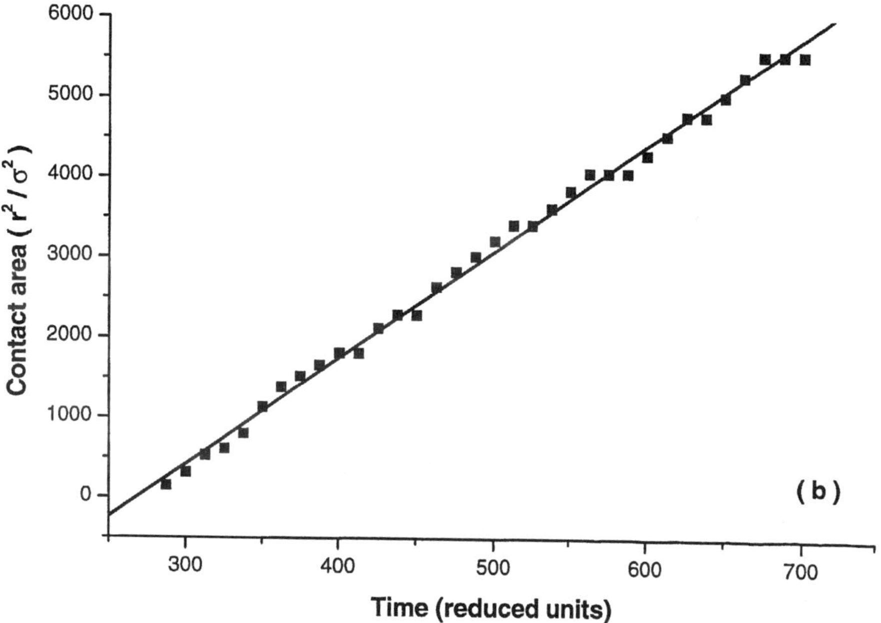

Figure 2. *Continued.*

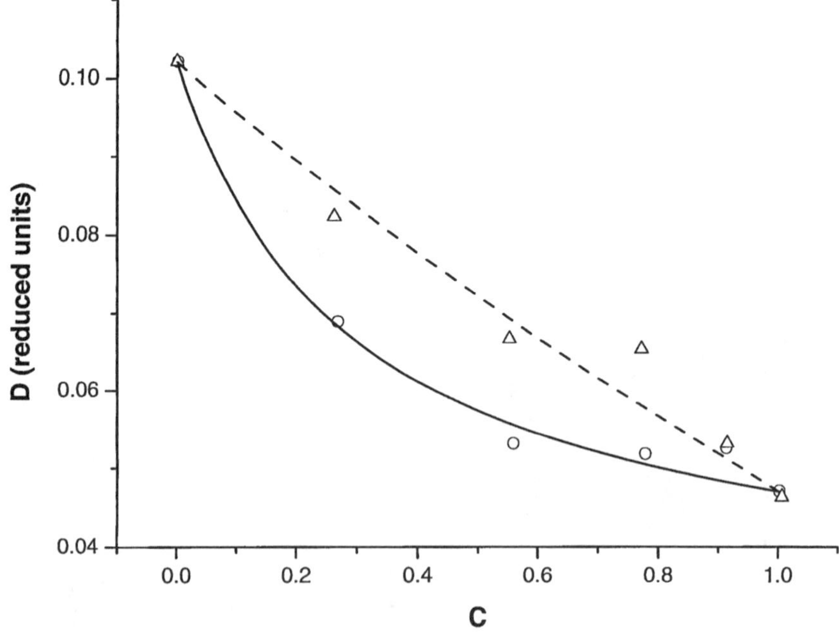

Figure 3. Influence of the heterogeneity concentration C on the diffusion coefficient of the liquid molecules (open circles : patterned substrates ; open triangles : random substrates). The plain and the dashed lines correspond to the best-fit results of equation 3 to the experimental data with respect to Λ.

less wettable species $C_{fs} = D_{fs} = 0.8$). The two sets of data can be successfully fitted by the equation 3. The best-fit values for Λ are 0.24 ± 0.04 for the patterned substrates and 0.82 ± 0.15 for the random ones. The fluctuations of the three phase contact line evidenced in Figure 1b and the comparison between the patterned and the random substrates suggest that the geometry of the heterogeneity patches (and not only their concentration) strongly influences the dynamics of spreading. Such substrates would behave differently from each other as a function of their respective percolation threshold (*1*).

The pseudo-diffusion process of the liquid molecules on the solution surface can be understood as the competition between the affinity of the liquid molecules for the substrate, which acts as a driving term, and a friction term (*10, 11*). This friction term can be interpreted as an Arrhenius process according to which ζ scales as

$$\zeta \cong \exp\left(\varepsilon_{fs}\right) \tag{6}$$

as a function of the solid/liquid energy of interaction ε_{fs} (*15*). D can therefore be expressed as :

$$D \cong A_1 \left(\varepsilon_{fs} - \varepsilon_{crit}\right) \exp\left(- A_2\, \varepsilon_{fs}\right) \tag{7}$$

where A_1 and A_2 are numerical constants and ε_{crit} is a threshold value of the solid/liquid interaction energy, which determines the wetting/non-wetting transition. The linear factor in ε_{fs} and the decreasing exponential respectively correspond to the driving term and inverse of the friction term. This can be understood as follows. For $\varepsilon_{fs} > \varepsilon_{crit}$, the linear term dominates at low ε_{fs} values (i.e. to an increase of the liquid/solid interaction corresponds an increase of the pseudo-diffusion coefficient D). At large values of ε_{fs}, the friction term dominates and D decreases even though the liquid/solid interaction is increased. In the cases of the simulations whose results are reported here, the main effect of the increase of the liquid/solid interactions is the increase of the driving term, and the subsequent increase of D (*15*).

Experimental results

To illustrate the interactions of a droplet on an heterogeneous substrate in the complete wetting regime, we experimentally considered the spreading of a low molecular weight silicon oil droplet on grafted silicon wafers.

Experimental techniques.

The silicon oil was a trimethyl-terminated poly(dimethylsiloxane) (viscosity $\eta = 0.010$ Pa s, surface tension $\gamma = 20.1$ mN/m), referred to as "PDMS10". Two kinds of heterogeneous substrates were prepared, using (100) silicon wafers as basis substrates.

Partial-OTS substrates.

Heterogeneous substrates partially grafted with octadecyltrichlorosilane (OTS) were obtained according to the procedure described in (*16-17*). These substrates will be referred to as the *partial-OTS substrates* (Figure 4a). The heterogeneities are prepared by varying the grafting time. Then the grafting procedure is carried out till completion, the obtained surface is a pure CH_3 surface. The mechanism of growth of these OTS monolayers has been extensively studied in the past few years (see e.g. *18-19* and the references therein). The surface of the so-obtained substrates consists in three types of regions, whose relative importance changes as a function of the dipping time of the substrate in the reaction solution.

1. The substrate surface is covered by a loose network of disordered OTS molecules forming an expanded liquid phase (LE*) so-called by analogy to the Langmuir layers. This network appears in the early times of the grafting process.

2. In parallel to the onset of this LE* phase, some islands of densely packed and self-organized OTS molecules appear (*18*). They are referred to as the condensed liquid phase (LC*) During the later stages of the grafting process, the LE* phase undergoes a phase transition and the structure of the coated layer spontaneously evolves to a completely organized structure characterized by the LC* phase.

3. The third region consists in the non-grafted parts of the substrate surface, i.e. silica regions bearing silanol groups at relatively low concentration. The critical surface tension γ_c of these surface components are in the following ranges : 20.5 to 21.0 mN/m for the LC* phase, 26.0 to 28.0 mN/m for the LE* phase and 28.0 to 30.0 mN/m for the silica surface. Due to the self-organization of the coating, the substrate becomes more and more hydrophobic. The critical surface tension γ_c of the substrates ranged from 20.5 to 24.5 mN/m.

Binary monolayers.

Secondly, we prepared binary self-avoiding monolayers (SAMs) from a two-component grafting solution containing saturated and unsaturated undecyltrichlorosilane (UTS and uUTS, resp.). These substrates will be referred to as the *UTS/uUTS substrate* (Figure 4b). The obtained surfaces range from pure CH_3 surfaces to pure CH_2 ones. For these substrates, the critical surface tensions determined from the corresponding Zisman plot obtained with the homologous series of alkanes ranged between 20.5 mN/m for the pure CH_3 surface (UTS) and 25.2 mN/m for the pure $CH= CH_2$ surface (uUTS).

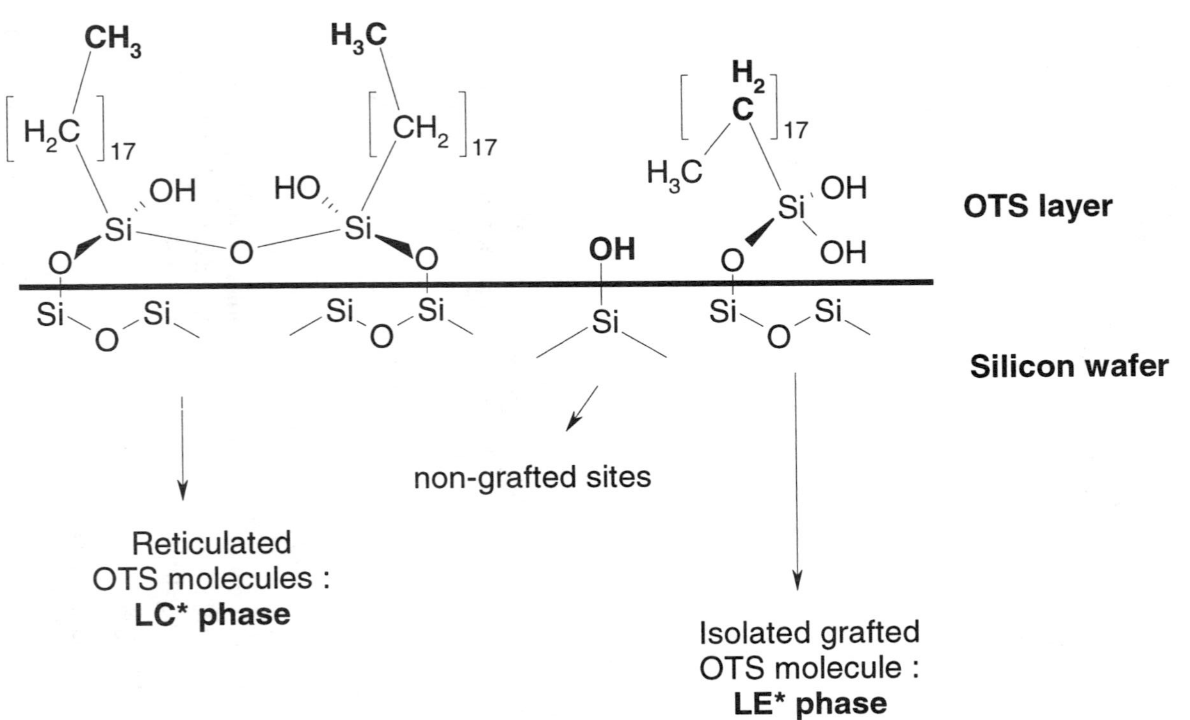

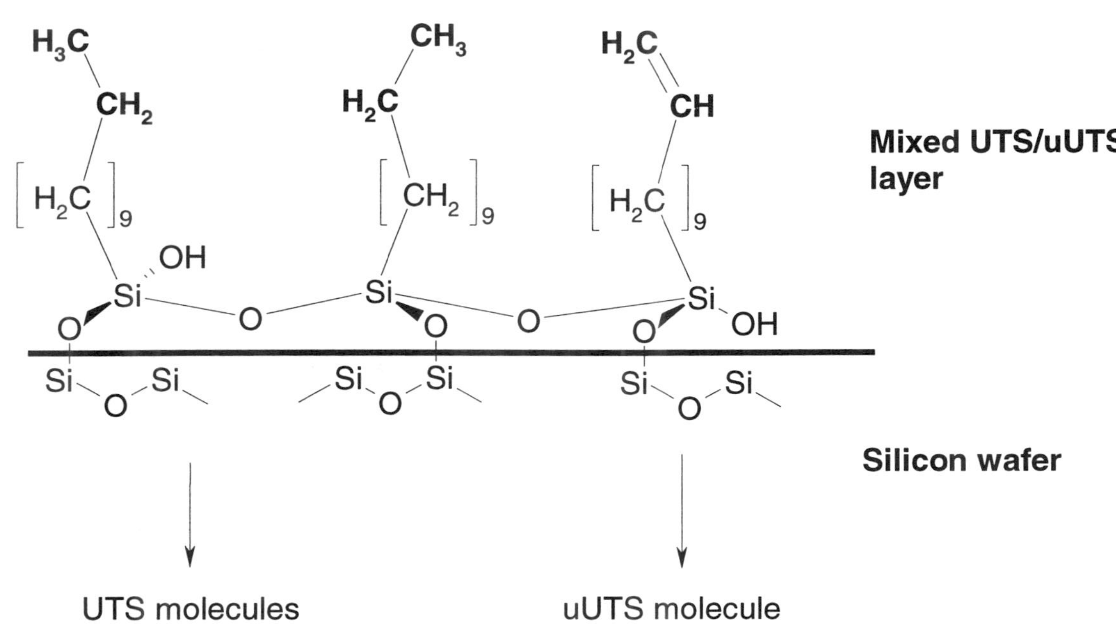

Besides the ranges of surface energies which are slightly different from each other, these two kinds of substrates mainly differ by the absence of physical roughness for the UTS/uUTS substrates, as explained below. To prepare partial-OTS substrates, the reaction time was varied to change the heterogeneity concentration at the wafer surface, while in the preparation of the UTS/uUTS surfaces, the grafting procedure was achieved till completion but the relative concentrations of the coating constituents were modified, in order to obtain a two-component SAM's. For these surfaces and as the length of both the UTS and uUTS molecules are equivalent, there is no physical roughness and the origin of the contact angle hysteresis has to be found in the molecular disorder.

For both kinds of substrates, the quality of the substrate coating has been estimated by spectroscopic ellipsometry, as proposed in (20), according to the effective medium approximation (21-22).

The dynamics of spreading of the PDMS10 droplets on top of these heterogeneous substrates has been investigated using high lateral resolution spectroscopic ellipsometry (HRSE), as reported in (14, 24). The lateral resolution of the SOPRA GESP 5 spectroellipsometer was 30 micrometers and the thickness resolution of the equipment was 0.02 nm. The incidence angle was 75 degrees, close to Brewter's angle of silicon. The ellipsometric angles were measured at 512 different wavelengths between 280 nm and 820 nm, using an intensified diode-array detector instead of a standard monochromator. After the deposition of the microdroplet on top of the substrate, the thickness of the liquid was measured every 80 micrometers by radially scanning the droplet at different times. Practically, the film length is calculated by subtracting from the base radius of the droplet, measured at mid-height of the film, the radius of the reservoir, measured 100 Å above the substrate baseline. When the edges of the film are rather smooth and that its thickness does not exceed 7 Å, the film radius is measured 3.5 Å above the baseline.

Droplet shapes on heterogeneous substrates.

Let us first consider the thickness profiles of the PDMS droplets spreading on these two kinds of substrates. As shown in Figure 5, the droplet shapes observed either on homogeneous or on heterogeneous substrates appears to be different from each other.

On the homogeneous SiO_2 (i.e. non-grafted) and OTS surfaces, a precursor film of molecular thickness progressively appears in front on the macroscopic part of the drop (Figure 5a and c). The thickness of this precursor film is either 7 Å in the former case (i.e. one layer of PDMS molecules laying flat on the substrate) and 14 Å in the latter. In both cases, the edges of the film are sharp, indicating that both surface heterogeneities and two dimensional volatility of the liquid are negligible. On the contrary, the effect of the surface heterogeneities clearly appears on the thickness

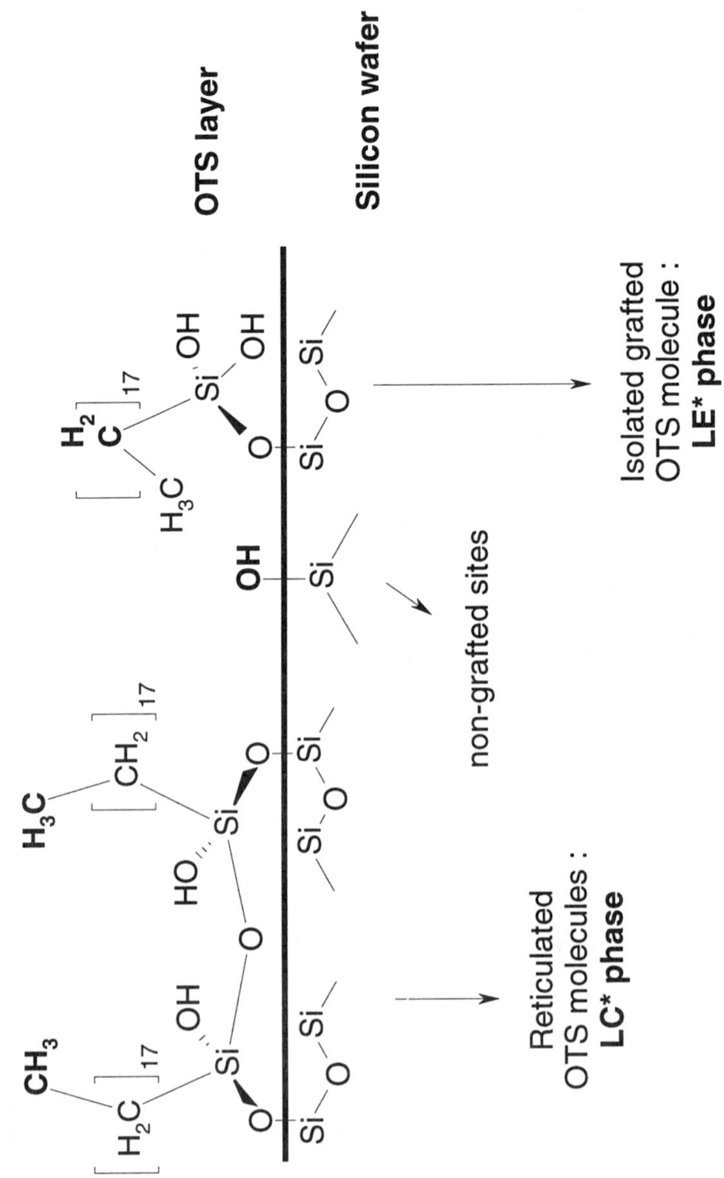

OTS layer
Silicon wafer
Isolated grafted
OTS molecule :
LE* phase
non-grafted sites
Reticulated
OTS molecules :
LC* phase
(a)

(b)

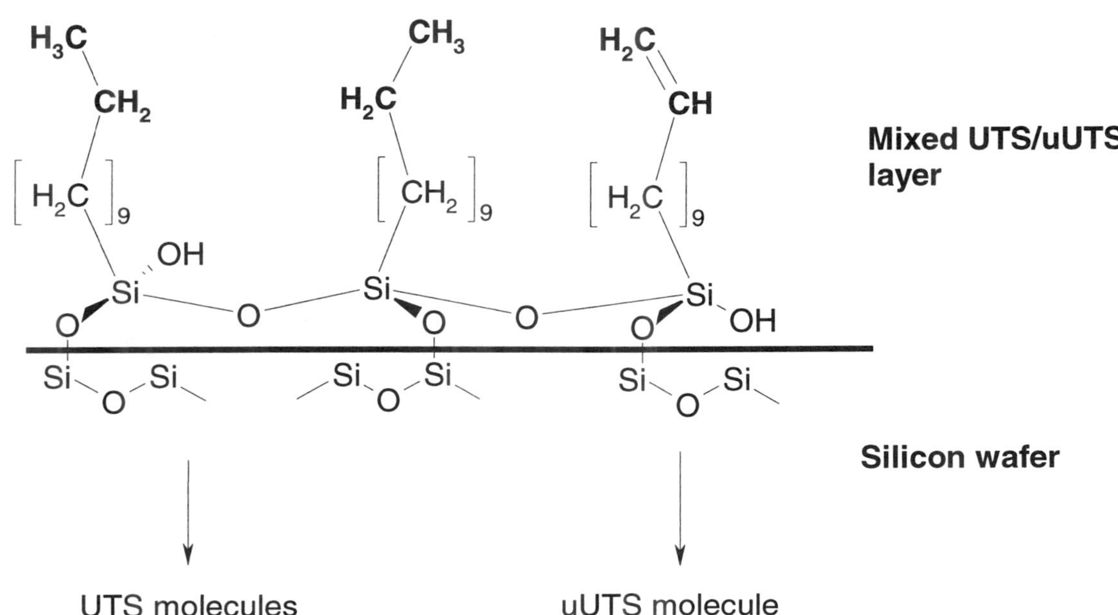

Figure 4. Heterogeneous substrates. (a) Partially OTS-grafted silicon wafers. (b) Mixed UTS/uUTS-grafted silicon wafers.

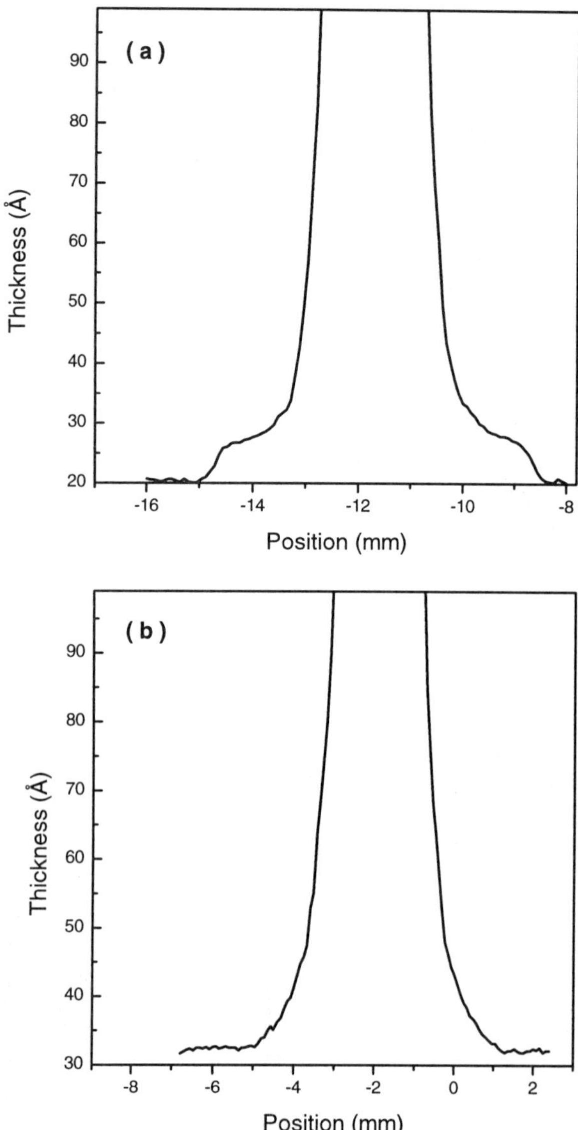

Figure 5. Thickness profiles of a PDMS10 microdroplet spreading on top of a solid substrate. (a) non-grafted. (b) 51 percent OTS-grafted surface. (c) Completely OTS-grafted wafer. (d) uUTS-grafted wafer. (e) 30 percent uUTS-grafted wafer. (f) 100 percent UTS-grafted wafer.

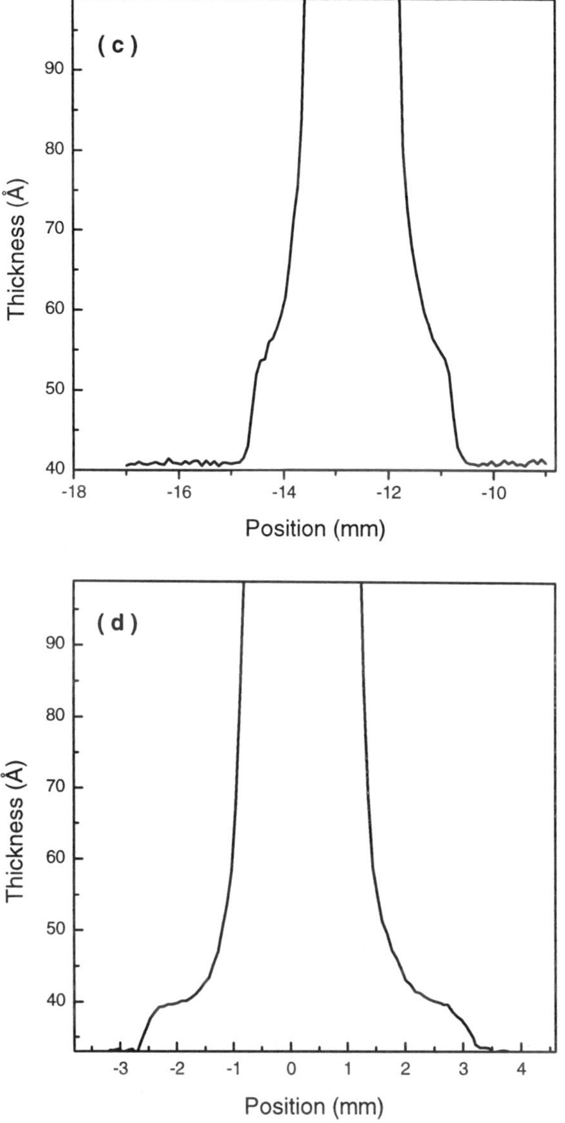

Figure 5. *Continued.*

Continued on next page.

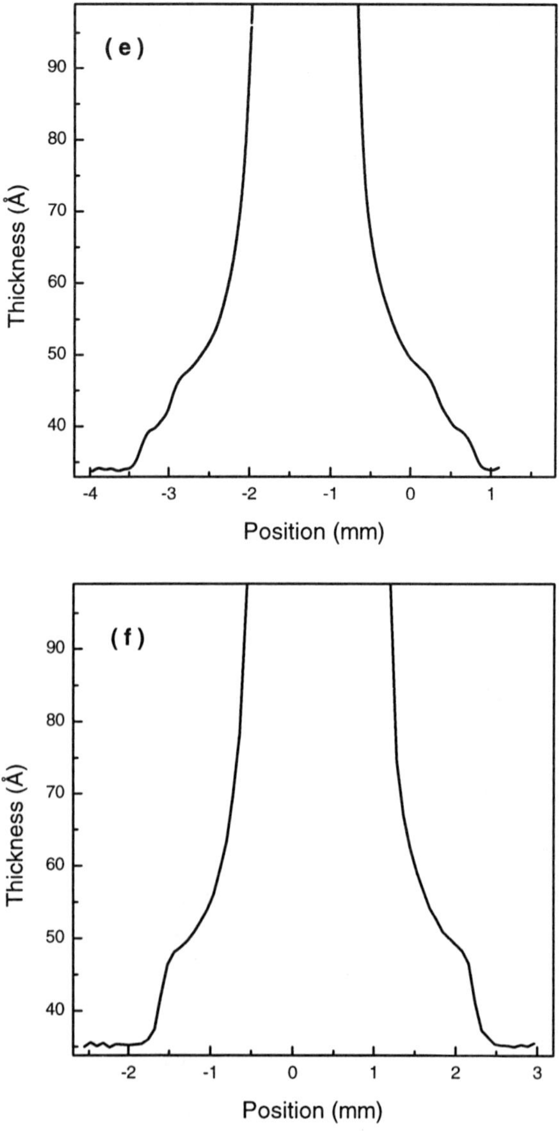

Figure 5. *Continued*

profiles measured from droplets spreading on partially grafted surfaces, as show on Figure 5b for 50 percent OTS-grafted substrates. The precursor film has no specific shape and its edges are rather smoother than in the previous cases.

Although, the shape of the droplet observed on the pure uUTS or UTS surfaces (Figure 5d and f) are respectively similar to that observed on the pure SiO_2 or OTS surfaces, the partially grafted substrates considered in both systems behave quite differently from each other, even at the same coating fraction. The droplet spreading on the 50 percent uUTS-grafted substrate (Figure 5e) exhibits a stratified pyramid shape. The thickness of each layer is equal to 7 Å, i.e. to the thickness of one PDMS molecule. This shape is similar to the one observed on substrates bearing of loosely grafted layer of trimethyl functional groups such as the one obtained by exposing a UV-ozone activated wafer to hexamethyldisilazane (HMDZ) vapors during 1 hr (7-9). It is also characteristic of a spreading process in which the energy dissipation is controlled by the liquid bulk viscosity and not by the friction of the liquid molecules on top of the solid substrate (24-25).

Diffusion coefficients.

By measuring the time evolution of the droplet thickness profiles, it is possible to measure the diffusion coefficient D of the liquid molecules associated to the spreading process. In all the cases reported in this study, the film length increases linearly with $t^{1/2}$, as expected (data not shown). The corresponding values of D are represented in Figure 6 as a function of C, the heterogeneity concentration on the corresponding substrate.

D monotonously decreases as the concentration of the less wettable molecules increases at the substrate surface. This decrease is correlated to the onset of the LC^* phase at the surface of the partial-OTS substrates. In that case, this behavior is clearly related to the patchy nature of the growth process and illustrates the MD simulations reported in the first part of this article. The experimental data are adequately fitted by our model (equation 3). The best-fit values for Λ are significantly different from each other and equal to 0.21 ± 0.02 and 0.41 ± 0.07 for the partial-OTS substrates and for the mixed UTS/uUTS surfaces, respectively. As expected for the results of the MD simulations, the increase of Λ could correspond to an increase of the randomness of the substrates.

Conclusions

The present studies brings experimental as well as computational evidences that, in the complete wetting regime, the diffusion coefficient D of liquid molecules on top of a heterogeneous substrate is a monotonous function of the heterogeneity concentration. Moreover, D can be expressed as a simple rational function of the

200

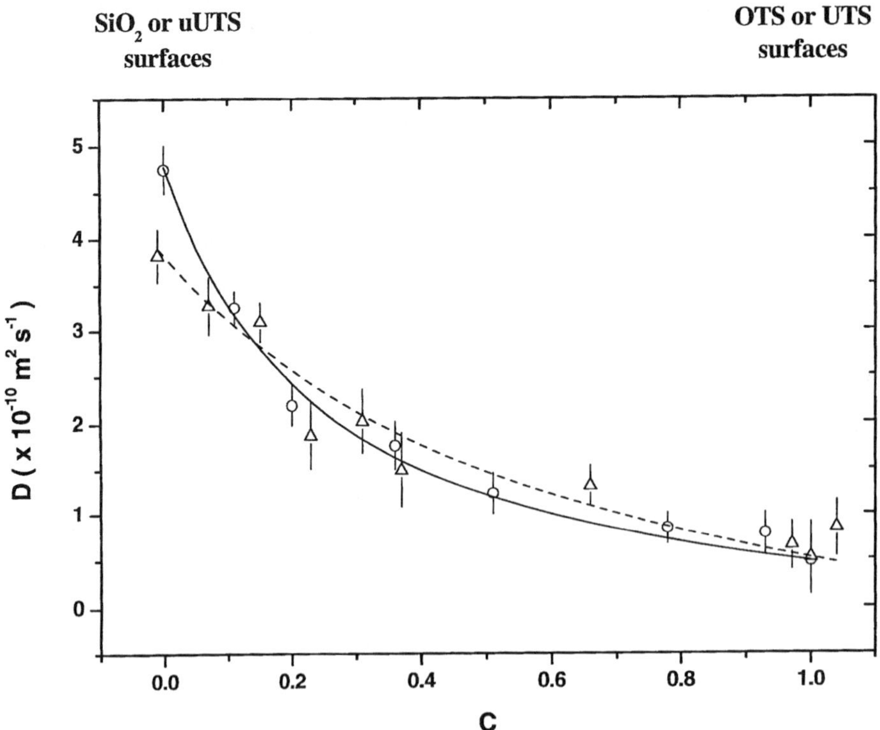

Figure 6. Influence of the heterogeneity concentration C *on the diffusion coefficient of the PDMS molecules (open circles : partially OTS-grafted substrates ; open triangles : mixed UTS/uUTS substrates). The plain and the dashed lines correspond to the best-fit results of equation 3 to the experimental data with respect to* Λ. *The values of* Λ *are respectively equal to 0.21 ± 0.02 and 0.41 ±0 .07 for these two types of surfaces.*

diffusion coefficients measured on the corresponding pure substrate, of the heterogeneity concentration and of a parameter Λ which implicitly accounts for the geometry of the heterogeneities. In a first approximation, Λ is equal to the ratio of the friction coefficients of the liquid molecules on the pure substrates ζ_A/ζ_B.

Acknowledgements.

This research is partially supported by the Ministère de la Région Wallonne.

References

1. de Gennes, P.G. *Rev. Mod. Phys.* **1985**, *57*, 827.
2. Collet, P.; De Coninck, J.; Dunlop, F.; Regnard, A. *Phys. Rev. Lett.* **1997**, *79*, 3704.
3. Adão, M.H.; de Ruijter, M.J.; Voué, M.; De Coninck, J. *Phys. Rev. E* **1999**, *59*, 746.
4. de Ruijter, M.J. Ph.D. thesis, Université de Mons-Hainaut, 1998.
5. Blake, T.D.; Haynes, J.M., *J. Coll. Interface Sci.* **1969** *30*, 421.
6. Blake, T.D., in *Wettability*; Berg, J.C., Ed.; Marcel Dekker: New-York, 1993.
7. Heslot, F.; Cazabat, A.M.; Levinson, P.; Fraysse, N. *Phys. Rev. Lett.* **1990**, *65*, 599.
8. Heslot, F.; Fraysse, N.; Cazabat, A.M. *Nature* **1989**, *338*, 640.
9. Heslot, F.; Cazabat, A.M.; Levinson, P. *Phys. Rev. Lett.* **1989**, *62*, 1286.
10. de Gennes, P.G.; Cazabat, A.M. *C. R. Acad. Sci.* **1990**, *310*, 1601.
11. Cazabat, A.M.; Valignat, M.P ; Villette, S.; De Coninck, J.; Louche, F. *Langmuir* **1997**, *13*, 4754.
12. De Coninck, J.; D'Ortona, U.; Koplick, J.; Banavar, J.R. *Phys. Rev. Lett.* **1995**, *74*, 928.
13. D'Ortona, U.; De Coninck, J.; Koplick, J., Banavar, J.R. *Phys. Rev. E* **1996**, *53*, 562.
14. Voué, M.; Semal, S.; De Coninck, J. *Langmuir* **1999,** *15*, 7855.
15. Voué, M.; Valignat, M.P.; Oshanin, G.; Cazabat, A.M.; De Coninck, J. *Langmuir* **1998,** *14*, 5951.
16. Broszka, J.B.; Shahidzadeh, N.; Rondelez, F. *Nature* **1992**, *360*, 719.
17. Broszka, J.B.; Ben Azouz, I.; Rondelez, F. *Langmuir* **1994**, *10*, 4367.
18. Davidovits, J.V. PhD thesis, Université Pierre et Marie Curie, Paris VI, 1998.
19. Schwartz, D.K.; Steinberg, S.; Israelachvili, J.; Zasadzinski, J.A.N. *Phys. Rev. Lett.* **1992,** *69*, 3354.
20. Semal, S.; Voué, M.; Dehuit, J.; de Ruijter, M.J.; De Coninck, J. *J. Phys. Chem. B* **1999**, *103*, 4854.
21. Azzam, R.M.A.; Bashara, N.M. *Ellipsometry and polarized Light*; North Holland: Amsterdam, 1977.

22. Aspnes, D.E.; Theeten, J.B.; Hottier, F. *Phys. Rev. B* **1979,** *20*, 3292.

23. Voué, M.; De Coninck, J.; Villette, S.; Valignat, M.P.; Cazabat, A.M. *Thin Solid Films* **1998**, *313-314*, 819.

24. Valignat, M.P.; Oshanin, G.; Villette, S.; Cazabat, A.M.; Moreau, M. *Phys. Rev. Lett.* **1998**, *80*, 5377.

25. Voué, M.; Valignat, M.P.; Oshanin, G.; Cazabat, A.M. *Langmuir* **1999**, 15, 1522.

26. De Ruijter, M.; Blake, T.D.; De Coninck, J. *Langmuir* **1999**, *15*, 7836.

Indexes

Author Index

Subject Index

D